A Comprehensive Review of the Versatile Dehydration Processes

Edited by Jelena D. Jovanović

Published in London, United Kingdom

A Comprehensive Review of the Versatile Dehydration Processes
http://dx.doi.org/10.5772/intechopen.100793
Edited by Jelena D. Jovanović

Contributors
Maria Inês Sucupira Maciel, Nathalia Barbosa da Silva, Patrícia Moreira Azoubel, Muhammad Faisal Manzoor, Sakhawat Riaz, Asifa Kabir, Aqsa Haroon, Anwar Ali, Sandeep Janghu, Bandita Bagchi Banerjee, Pham Van Kien, Nguyen Hay, Le Quang Huy, Jelena D. Jovanović, Borivoj K. Adnadjevic

First published in London, United Kingdom, 2023 by IntechOpen
IntechOpen is the global imprint of INTECHOPEN LIMITED, registered in England and Wales, registration number: 11086078, 5 Princes Gate Court, London, SW7 2QJ, United Kingdom

British Library Cataloguing-in-Publication Data
A catalogue record for this book is available from the British Library

Additional hard and PDF copies can be obtained from orders@intechopen.com

A Comprehensive Review of the Versatile Dehydration Processes
Edited by Jelena D. Jovanović
p. cm.
Print ISBN 978-1-83768-140-2
Online ISBN 978-1-83768-141-9
eBook (PDF) ISBN 978-1-83768-142-6

Meet the editor

Dr. Jelena D. Jovanović is a research professor at the Institute of General and Physical Chemistry, Belgrade, Serbia. Her interests include advanced and smart materials, polymers, composites, hydrogels, poly(siloxanes), synthesis and polymerizations, and physicochemical processes (adsorption, extraction, swelling, dehydration, drug-release) under both conventional and non-conventional conditions (microwaves, ultrasonic, cavitation). Dr. Jovanovic is also interested in hydrogels and the effects of external fields on reaction kinetics. She has worked on novel methods of kinetics analysis. Dr. Jovanović has extensive experience working on international projects and has stayed abroad several times. She is a reviewer for international journals and an external expert for the European Commission for proposal evaluations.

Contents

Preface

This book discusses dehydration processes. Chapter 1 examines the significance of water and dehydration processes in general. It reviews the two dehydration processes of food dehydration and hydrogel dehydration. It presents the physical basics of dehydration and discusses the state of water in food and hydrogels. Regarding food dehydration, the chapter demonstrates the significance of controlling the structural-kinetic states of water and their effect on the structural and sensing properties of dehydrated food. Regarding hydrogel dehydration, the chapter discusses its practical and theoretical importance and suitability for modeling the kinetics of food dehydration. The chapter also presents novel kinetic models that can describe hydrogel dehydration with great precision and reliability.

Chapter 2 presents recent developments in energy-efficient drying technologies for food dehydration. The authors discuss conventional dehydration processes based on solar dehydration, trash drying, smoke, drum, spray dehydration, fluidized bed drying, and freeze drying. They also describe advanced drying processes that include infrared, microwave, ultrasonic and other non-conventional drying techniques. Innovative food dehydrating methods use less energy and therefore contribute to environmental protection. During the dehydration process, some changes in dried food products may occur. Novel dehydration methods better conserve the chemical structure, color, taste, flavor, and appearance of the dried product. The chapter shows that the various dehydration techniques do not affect the fundamental structure of polysaccharides and that certain foods subjected to dehydration processes exhibit increased total phenolic content.

Chapter 3 reviews the state of the art in dehydration assisted by microwave (MW), infrared radiation (IR), and radio frequency (RF) applied to food drying, specifically to foods such as bananas and apples. It discusses the advantages and disadvantages of each method compared to conventional drying systems. IR dehydration in comparison to conventional dehydration exhibits certain advantages, such as a higher heating rate, shorter drying time, and greater quality of the dried product. As such, IR drying techniques are increasingly being used method for drying food, vegetables, grains, fruits, and other high-value products. MW has also been widely applied in the drying of various foods and is a proven method for improving the drying process and quality of dried products. RF has also been studied and applied in food processing. Each of the methods discussed in this chapter can be combined with other drying methods, such as hot air drying, heat pump drying, vacuum drying, and freeze drying.

Chapter 4 presents the theoretical basis of hybrid drying techniques based on MW drying, which can be enhanced via combination with other drying methods such as hot air drying, freeze drying, vacuum drying, and fluidized bed drying. Using MW in hybrid drying significantly enhances drying rates, making it a novel approach to retaining the quality of dried products. Each hybrid drying method has its own advantages. For example, hot air drying speeds up the removal of moisture from the core of the food to the surface, whereas freeze drying assists in preserving the bioactive compounds and nutritional status of products. Fluidized bed drying and vacuum drying are preferable techniques in cases when a better rehydration ratio and uniform heating of the products are required. The chapter describes why the holistic approach is crucial to developing smart hybrid drying systems that bring together the efficiency of drying and the quality of the dried products.

Chapter 5 examines the effects of pre-treatment by ultrasound (US) in different media (water and ethanol) in the convective drying of BRS Vitória grapes. The study shows that pre-treatment with the US increased the efficiency of convective drying of the grapes by reducing drying time up to 61% when using ethanol as media. The pre-treatment did not result in any significant effects of media on the texture, color, soluble solids, and water activity of the grapes. Pre-treatment with ethanol is thus revealed as effective in obtaining raisins, reducing drying time, maintaining the quality of the product, and promoting more retention of nutrients. There was no observed loss of phenolic content in grapes after drying. US combined with ethanol exhibited the highest phenolic content of the treatments. Regarding the kinetics of the investigated drying process, the chapter study establishes that the logarithm model is the best to describe the kinetics of the grape drying process when compared to other mathematical models.

Finally, Chapter 6 summarizes investigations in the kinetics of absorbed water dehydration from different hydrogels, including equilibrium swollen hydrogels of poly(acrylic acid) hydrogel (PAAH), poly(acrylic-co-methacrylic acid) (PAMAH), and poly(acrylic acid)-g-gelatin (PAAGH). The complex kinetics of dehydration of hydrogels are described by a series of novel kinetic models: distribution apparent energy activation model (DAEM), Webull's distribution of reaction times, the dependence of the degree of conversion (α) on the temperature defined by the logistic function, coupled single step-approximation, and iso-conversion curve. The chapter explains the complex kinetics of dehydration of hydrogels in terms of the fluctuating structure of the hydrogel, the phase state of the absorbed water, and the thermal activation of the hydrogel.

Jelena D. Jovanović
Research Professor,
Institute of General and Physical Chemistry,
Belgrade, Serbia

Chapter 1

Introductory Chapter: A Comprehensive Review of the Versatile Dehydration Processes

Jelena D. Jovanović and Borivoj K. Adnadjević

1. Introduction

Water is the most abundant substance on the Earth and the main component of plant and animal tissues, in which it plays a role as a solvent and a reagent. The unique role of water in natural processes is related to its physical and chemical properties and its widespread. As a result, most materials in the natural conditions contain water either as chemically bound or retained in pores due to intermolecular interactions [1]. The presence of water in food and foodstuffs plays a significant role in the physico-chemical and biological processes that take place during their storage [2].

Dehydration is a complex reversible and endothermic physicochemical process of removing water from the material, which takes place under conditions of simulated energy exchange (especially heat) and mass transfer between the material and the external environment [3]. The removal of water is a kinetically complex process characterized by either rapid nucleation of water molecules at the reaction boundary phase (RBP) or nucleation at certain locations of the boundary phase (RBP), after which there is an increase in the size of the nucleus, which leads to the removal of water from the material. The most important feature of the dehydration process is the dominant influence of the dehydration product on the mechanism and kinetics of water removal from the material [4].

2. Food dehydration

The water content in food, fruits, vegetables, foodstuffs, and in agricultural products varies in a wide range from 60 to 98% by mass. The dominant content of water indicates the key role of water on the physicochemical, biological, and nutritional and sensing properties of the food. Water is, first of all, the most important medium in which chemicals (ions, salts, vitamins, etc.,) and biological reagents (sugars, proteins, lipids, DNA, enzymes, etc.,) move, collide, and react. In addition, water participates such as (a) reagent and coreagent in a series of degradation reactions (hydrolysis of lipids, Maillard reactions, enzymatic browning, vitamin degradation, etc.,); (b) stabilizes the most important biological structures, (enzymes, proteins, DNA, and cellular membrane); (c) control the growth of pathogens and other microorganisms; (d) significantly changes the physical and chemical properties of the material (thermal conductivity, thermal capacity, electrical and dielectric properties), etc. [2]

IntechOpen

In regard to that, it is unambiguous that state of water and its physical–chemical (structural–kinetic) during the dehydration of the material has the key influence on dehydration process, chemical and biological reactions, and nutritional and sensory properties of food. Water in food and foodstuffs exists in three different structural-kinetic states, namely bond water, intermediate, and freeze water. Bound water is formed as a consequence of the formation of hydrogen bonds between water molecules and polar groups on RBP. Free water is formed by a mutual interaction between water molecules and is similar in structure to bulk water. Intermediate water is formed from water molecules that interact weekly with RBP. Different structural-kinetic states of water lead to different interactions between water molecules and chemical components of food [5]. In order to understand and govern food dehydration process, it is necessary to focus further research on expanding and deepening the knowledge to (a) structural-kinetic state of water in food, (b) change in the structural-kinetic state of water due to dehydration; (c) interaction of different states of water with chemical and biological components of food.

The term "drying" is a synonym for food dehydration by application of heat and is the oldest method for preserving food. The main reason for drying food is to extend the shelf life of fresh materials without the use of cooling and storage, because the reduction of water content inhibits growth and the development of spoilage and pathogens microorganism reduces the activity of enzymes and reduces unwanted degradation reactions. The process of drying food also leads to a reduction in the weight and volume of food, which significantly affects the costs of packaging, storage, and transportation of food. Drying food leads to a change in color, texture, and smell compared to fresh material and a reduction in the nutritional value of food [6].

Drying is an energy-consuming process and the cost of used energy compared to other storage methods is relatively high with predictable growth for the near future. Accordingly, with aim to reducing specific energy consumption of drying and obtaining a product with preserved nutritional and sensory properties, a number of conventional (sun, hot-air, spray-draying, freeze-drying, fluidized-bed drying, and osmotic dehydration) and innovative (microwave drying, infrared drying, solar drying, electric and magnetic field dewatering, and ultrasonic dehydration) food-drying methods have been developed. In order to improve the existing technological processes of food drying and developing new ones, further research should be focused on the (a) development and advancement of new processes of uniform volume of heating of food; (b) determination and control of the physical-chemical state of water molecules inside the tissue during dehydration and the ones that move inside the material when leaving; (c) understanding and managing of the process of changing the state of the matrix during dehydration.

3. Kinetics models of hydrogel dehydration

Hydrogels are mainly defined as three-dimensional, cross-linked hydrophilic polymeric networks which have the ability to absorb a significant amount of water or other aqueous fluids (swelling) without dissolving or losing structural integrity [7]. Hydrogels are extremely prominent against other polymeric materials because of their characteristic properties such as smart response to external stimuli, swelling ability, high water content, biocompatibility, adjustable porosity, and mechanical properties. The most outstanding are their high swelling capacity and resemblance to living tissues more than any other type of artificial biomaterials. Because of these distinguishing properties, hydrogels have been widely used in versatile applications from biomedical to green energy. Their use are mostly recognizable in pharmacy, medicine, and biomedical applications [8, 9], especially in controlled and targeted drug release [10], regenerative medicine and tissue engineering, contact lenses, biosensors, etc. [11]. Hydrogels are excellent for applications in biotechnology, environmental protection, agrochemistry, horticulture, cosmetics, as superabsorbents in hygiene products, packaging materials for food storage, in textile materials, in sensor materials, etc. In recent times, the applications of hydrogels and hydrogel-derived materials present novel materials for electrochemical energy conversion systems due to their specific and tailorable physicochemical properties [12].

Due to high water content, hydrogels should be assumed as model systems suitable for modeling the description of the kinetics of food dehydration. Similarly, as in food, water in hydrogels can be classified generally into three types: (a) bound water, which involves strongly bound and weakly bound; (b) associated water, involving strongly associated and weakly associated water; and (c) free water. According to their phase transition behavior, three types of water in hydrogels have been identified: nonfreezing, freezing bound, and free non-bound water [13]. Many physical properties of hydrogels depend on the organization of water within and at the surface of hydrogels [14]. The structures of the polymer network and the embedded water are important factors governing the physicochemical properties of hydrogel materials [15].

Knowledge and governing of the hydrogel dehydration process is of extraordinary practical and theoretical importance. In the literature, the kinetics of dehydration of hydrogels is most often described by the diffusion kinetic model [16]. However, new kinetic models have been developed that can describe hydrogel dehydration more precisely and with a higher degree of reliability. The complex kinetics of dehydration of hydrogels was described by a series of novel kinetic models: distribution apparent energy activation model, Webull's distribution of reaction times, the dependence of the degree of conversion (α) on the temperature which is defined by the logistic function, coupled single step-approximation with iso-conversional curve. These models were applied to evaluate the dehydration kinetics of different hydrogels: poly(acrylic acid) hydrogel, poly(acrylic-co-methacrylic acid), and poly(acrylic acid)-g-gelatin. It was determined that these new kinetic models can very appropriately describe the kinetics of dehydration of the investigated hydrogels covering the whole range of the dehydration process. The correlations among the values of the rate constants (k), activation energy (Ea), preexponential factor (lnA), with the primary structural properties of the investigated xerogels (hydrogels in dry state) were determined [17–22].

4. Conclusions

Understanding and governing the possibility of controlling the structural–kinetic states of water in food has a key role (key-role) in the mechanism and kinetics of food dehydration and preservation.

Dehydration is one of the most important operations in food science. Dehydration enables extension of shelf life and preservation of their physicochemical, biological, nutritional, and sensing properties. Understanding and managing the possibility of controlling the structural-kinetic states of water in food has a key role (key-role) in the mechanism and kinetics of food dehydration and the preservation of physicochemical, biological, and nutritional and sensing properties.

Due to their unique properties, hydrogels found versatile application. Knowledge and governing of the hydrogel dehydration process is of extraordinary practical and theoretical importance. Hydrogels are suitable for modeling the kinetics of food dehydration.

Author details

Jelena D. Jovanović[1*] and Borivoj K. Adnadjević[2]

1 Institute of General and Physical Chemistry, Belgrade, Serbia

2 Faculty of Physical Chemistry, University of Belgrade, Belgrade, Serbia

*Address all correspondence to: jelenajov2000@yahoo.com

References

[1] Lewicki PP. Water effects on physical properties of raw materials and foods. In: Gliński J, Horabik J, Lipiec J, editors. Encyclopedia of Agrophysics. Encyclopedia of Earth Sciences Series. Dordrecht: Springer; 2011. DOI: 10.1007/978-90-481-3585-1_180

[2] Derossi A, Cassi D, Severini C. Mass transfer mechanisms during dehydration of vegetable food: Traditional and innovative approaches. In: Adv. Topic in Mass Transfer, Chapter 15. London, UK, London, UK: IntechOpen; 2011. pp. 304-320

[3] Adnadjevic B, Jovanovic J. HYDROGELS – Synthesis, Structure and Properties. Belgrade: Faculty of Physical Chemistry; 2014. pp. 1-253. ISBN: 978-86-82139-46-1

[4] Brown WE, Dollimore D, Galwey AK. Reactions in the solid state. In: Bamford CH, Tipper FH, editors. Comprehensive Chemical Kinetics. Vol. 22. Amsterdam: Elsevier; 1980. pp. 1-340

[5] Jhon MS, Andrade JD. Water and hydrogels. Journal of Biomedical Materials Research. 1973;7:509-522

[6] Onyeaka HN, Nwabor OF. Food ecology and microbial food spoilage, chapter 2. In: Onyeaka HN, Nwabor OF, editors. Food Preservation and Safety of Natural Products. Academic Press; 2022. pp. 3-18, ISBN: 9780323857000. DOI: 10.1016/B978 0-323-85700-0.00018-6

[7] Peppas NA, Slaughter BV, Kanzelberger MA. 9.20—Hydrogels. In: Matyjaszewski K, Möller M, editors. Polymer Science: A Comprehensive Reference. Amsterdam: Elsevier; 2012. pp. 385-395. DOI: 10.1016/B978-0-444-53349-4.00226-0

[8] Peppas NA, Bures P, Leobandung W, Ichikawa H. Hydrogels in pharmaceutical formulations. European Journal of Pharmaceutics and Biopharmaceutics. 2000;**50**(1):27-46. DOI: 10.1016/S0939-6411(00)00090-4

[9] Caló E, Khutoryanskiy VV. Biomedical applications of hydrogels: A review of patents and commercial products. European Polymer Journal. 2015;**65**:252-267. DOI: 10.1016/j.eurpolymj.2014.11.024

[10] Peppas NA. Hydrogels and drug delivery. Current Opinion in Colloid & Interface Science. 1997;**2**(5):531-537. DOI: 10.1016/S1359-0294(97)80103-3

[11] Alma TB, Chigozie AN, Emmanuel MB. Recent advances in biosensing in tissue engineering and regenerative medicine. In: Vahid A, Selcan K, editors. Biosignal Processing. Rijeka: IntechOpen; 2022, p. Ch. 15. DOI: 10.5772/intechopen.104922

[12] Guo Y, Bae J, Fang Z, Li P, Zhao F, Yu G. Hydrogels and hydrogel-derived materials for energy and water sustainability. Chemical Reviews. 2020;**120**(15):7642-7707. DOI: 10.1021/acs.chemrev.0c00345

[13] Gun'ko VM, Savina IN, Mikhalovsky SV. Properties of water bound in hydrogels, Review. Gels. 2017;**3**:37. DOI: 10.3390/gels3040037

[14] Yoshida H, Hatakeyama T, Hatakeyama H. Characterization of water in polysaccharide hydrogels by DSC. Journal of Thermal Analysis. 1992;**40**:483-489

[15] Naohara R, Narita K, Ikeda-Fukazawa T. Change in hydrogen bonding structures of a hydrogel with dehydration. Chemical Physics Letters. 2017;**670**:84-88

[16] Siepmann J, Siepmann F. Modeling of diffusion controlled drug delivery. Journal of Controlled Release. 2012;**161**:351-382

[17] Jovanović J, Adnadjević B, Ostojić S, Kićanović M. An investigation of the dehydration of Superabsorbing polyacrylic hydrogels. Materials Science Forum. 2004;**453-454**:543-548. DOI: 10.4028/www.scientific.net/MSF.453-454.543

[18] Adnađević B, Janković B, Kolar-Anić L, Minić D. Normalized Weibull distribution function for modelling the kinetics of non-isothermal dehydration of equilibrium swollen poly(acrylic acid) hydrogel. Chemical Engineering Journal. 2007;**130**(1):11-17. DOI: 10.1016/j.cej.2006.11.007

[19] Adnađević B, Janković B, Kolar-Anić L, Jovanović J. Application of the Weibull distribution function for modeling the kinetics of isothermal dehydration of equilibrium swollen poly (acrylic acid) hydrogel. Reactive and Functional Polymers. 2009;**69**(3):151-158. DOI: 10.1016/j.reactfunctpolym.2008.12.011

[20] Adnadjevic B, Tasic G, Jovanovic J. Kinetic of non-isothermal dehydration of equilibrium swollen poly(acrylic acid-co-methacrylic acid) hydrogel. Thermochimica Acta. 2011;**512**(1):157-162. DOI: 10.1016/j.tca.2010.09.019

[21] Adnadjevic B, Jovanovic J, Micic U. The kinetics of isothermal dehydration of equilibrium swollen hydrogel of poly(acrylic-co-methacrylic acid). Chemical Industry. 2009;**63**(5):585-591. DOI: 10.2298/hemind0905585a

[22] Adnadjević B, Gigov M, Jovanović J. Comparative analyses on isothermal kinetics of water evaporation and PAAG hydrogel dehydration under the microwave heating conditions. Chemical Engineering Research and Design. 2017;**122**:113-120. DOI: 10.1016/j.cherd.2017.04.009

Chapter 2

Food Dehydration Recent Advances and Approaches

Sakhawat Riaz, Asifa Kabir, Aqsa Haroon, Anwar Ali and Muhammad Faisal Manzoor

Abstract

Dehydration of organic material is undoubtedly a controlled attempt to conserve or construct a novel construct that will satisfy functional devotions. Food dehydration is reviewed in light of the latest progress in food materials research. Understanding the mechanics behind the drying process is crucial in food and agricultural product dehydration. Among the most crucial steps in preserving food is dehydration. Food drying innovations include photovoltaic, thermal imaging, microwave-assisted, and comparable hybrid technologies. According to a recent study, unique food dehydration technologies might increase drying efficiency by decreasing energy usage while improving product quality. Unique drying methods reduce food component degradation and create novel items for customers. Each method's use of specific foods will be reviewed in this chapter.

Keywords: food dehydration, novel technologies, benefits, application, food product

1. Introduction

Dehydrating is essential in many agricultural, food, biotechnological, mineral processing, pulp, wood, polymer, ceramics, pharmaceutical, paper, and chemical applications [1]. Perhaps dehydration is chemical science's oldest and most versatile way of drying procedures [2]. The fruit has many valuable chemicals, making it an essential element of the human diet [3]. Since the moisture content of fresh fruits exceeds 80%, they are considered perishable foods [4]. Around 40% of postharvest losses account for total fruit output in developing nations, such as India, significantly reducing the availability of fruits to customers [5]. During the entire seasons, fruits are gathered, but due to an absence of storage [6] and preservation accommodations, the marketplaces get congested, and the fruits start decaying before the end buyers can reach them [7]. During the drying process, reduction in the moisture content and, therefore, water activity permits the microbial activity in food materials to stabilize while also controlling supplementary deteriorative processes, such as browning, enzymatic and nonenzymatic reactions, lipid oxidation, and many more. Dehydrating food helps prevent bacterial growth that causes changes in chemicals and the occurrence of spoilage and the in food by reducing the moisture content of dietary items [8]. There are several goals for dehydrating dietary products. The most obvious is food preservation by dehydration. The dehydration technique limits microbial activity and other effects by lowering the humidity level of the item [9]. This method not only preserves the food from a microbiological aspect but also preserves its flavor and nutritional properties. The process of removal of water with heat is defined as dehydration. The earliest known method of food preservation was probably dehydration [10]. Fruits can also be sun-dried, and fish and meat can be smoked using well-known traditional methods [11]. A dehydrated food item has the benefit of being lightweight, which reduces shipping costs. However, the quality of the dried product is frequently diminished because high temperature is required in most conventional drying processes [12]. Many alternative approaches must be considered for potential application in the food industry [13, 14]. Vegetable drying was first recorded in the eighteenth century. Following that [15], scenarios of a world war were inextricably linked to the expansion of the drying industry [13]. During the Crimean War (1854–1856), British troops received dry vegetables from home; during the Boer War (1899–1902), Canada sent dried vegetables to South Africa; and during World War 1, the United States shipped about 4500 tons of dehydrated vegetables [16]. Potatoes, cabbage, spinach, turnips, carrots, celery, sweet corn, green beans, and soup mixtures were among the products processed in the United States [8].

2. Dehydration of food and food products

Keeping food fresh is the best way to protect its nutritional value, but most storage procedures call for relatively lower temperatures, which are unable to preserve all over the supply chain [8]. Preprocessing techniques, such as osmotic dehydration, before freeze drying, may partially remove the water before the final drying stage [17]. When choosing a dryer, consider the material's physical features, manufacturing capability, initial moisture content, particle size distribution, drying attributes, maximum permissible product heating rate, and explosion character traits (such as spray or fluid bed dryings) [18]. When dehydrating food, it is essential to consider factors, such as moisture content, glass transition temperature, dehydration methodologies and hypotheses, and physical and chemical changes [19]. The drying process affects the product's chemical and physical characteristics and water content. Some characteristics that have been utilized to classify dried foods include water activity, isotherms, sorption, deterioration of microbes, enzymatic and nonenzymatic reactions, structural and physical phenomena, and degeneration of nutrient levels, perfumery, and tastes [20, 21].

Moisture diffusion influences the process of drying during the rate of the first falling stage. According to Fick's second law of diffusion, the tortuosity, intercellular space, and distortion of tissues of vegetables, as well as the structure and chemical composition of the food, all impact moisture transport during the dehydration process [22]. It is crucial to examine the compatibility of intrinsic barriers and how to alter them given the number of food items that have undergone evaporation and dehydration in command to maximize the benefits of drying and dehydration [23]. High latent heat of vaporization of water is the primary method of removing water from food was climate change. Climate change needs much energy, for example, 2.26 MJ/100°C [24]. Mechanical pressing was commonly employed before thermal drying to remove 20–30% of the water from solid food wastes. The variety of food drying procedures and equipment demonstrates the complexity of handling and processing solid foods. As well as the unique criteria for various food products.

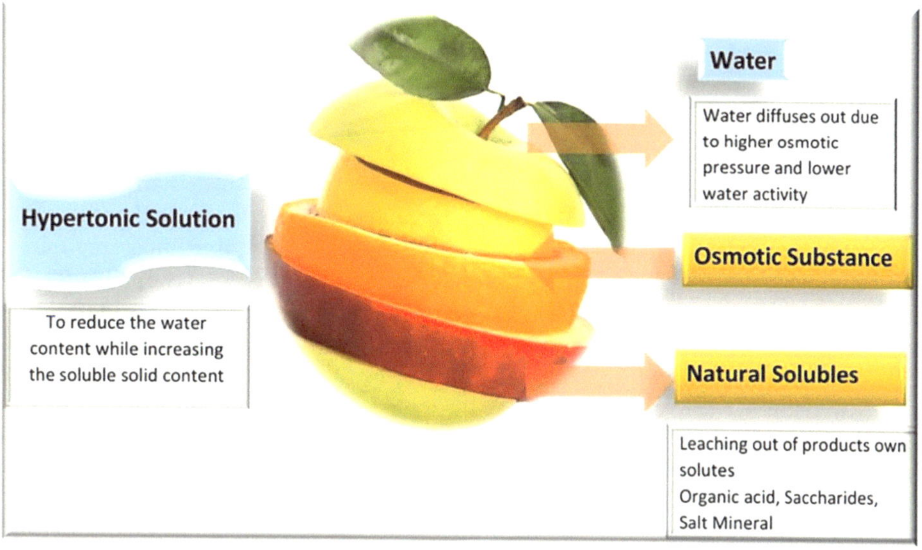

Figure 1.
Schematic demonstration of the osmotic dehydration process.

Furthermore, economic issues are a big concern, given the high amount of low-cost goods, such as skim milk [25].

Some water may be kept together by interactions between water molecules, resulting in a multilayer of water molecules. At the same temperature, this type of water, referred to as "bound water," has a lower partial pressure than pure water [26]. Additionally, the heat of vaporization for bound water is higher than pure water during similar temperatures. A food product's chemical composition directly affects how much bound water is present in it [27]. A product retains free moisture over its equilibrium moisture content. Only free moisture in a product may be eliminated during a specific dehydration technique [28]. The product type, temperature, and concentration of water vapor in the air influence the free moisture content of a product [28]. The bonded water in a food product is removed with considerable effort during the dehydration process [28]. Dried foods have various advantages, including improved storage stability, reduced packing requirements, and reduced transit bulk. **Figure 1** shows the osmotic dehydration process in food.

3. Conventional food dehydration processes

3.1 Solar dehydration

Between the end of the 1800s and the start of the 1990s, artificial dehydration took the place of sun dehydration [29]. Vegetables were first mechanically dried in the eighteenth century, an improvement over solar drying. It is a monitored and efficient solar energy system. Solar driers may produce hotter air and lower relative humidity [30]. The oldest industrial technique still in use is probably this one. It has been used with various things since antiquity, comprising meat, fruits, plants, and fish [31, 32]. However, this strategy significant downsides bound its application to industrial manufacturing. Among all are the prerequisites for vast amounts of heavy labor inputs and space, the challenge of monitoring the drying frequency, pest infestations, and microbial contamination [33].

3.2 Tray drying

It has a simple design and can dry much stuff. The first hot air dehydrator was invented in 1795 and used to dry fruit and vegetables, such as raisins and prunes [29]. The proper operation of the tray dryer depends on an even distribution of airflow over the trays [29]. Colak and Hepbasli developed a green olive model in 2007. The energy efficiency first study of dehydrators in 1921 by Christie and Cruss [34]. At the time, heated forced-air dehydrators were used to dry prunes instead of sun use [8]. The prime problem of the tray dryer is irregular drying brought on by inadequate airflow dispersion in the drying chamber [29]. The efficiency of a tray dryer system can be increased and drying nonuniformity minimized. Due to the systems' low operating costs, many dryer structures have been created using solar energy [8].

3.3 Smoke dehydrating

To preserve food through smoking is almost as ancient as direct-air dehydration. The two strategies are frequently employed in tandem [35]. Smoke has the additional benefit of giving attractive tastes to meals [5]. Furthermore, some of the chemicals produced by smoking have antibacterial characteristics [36]. While primarily not employed to lower the food content of moisture, the heat involved in smoke production has a drying impact [36]. Smoking always associates with fish and meat.

3.4 Drum dehydrating

Drum drying began 120 years ago, in the early 1900s, with Just Hatmaker developing the first drum dryer in 1902 [29]. Initially, a double drum dryer was created with feed going into the nip. It was less suitable for viscous liquids, so in 1945 a single barrel with top feed was developed to handle viscous products [29]. Feed was applied using dipping, splashing, spraying, and bottom feed rolls in a single drum. The feed's viscosity typically dictates the feeding method [37]. Food dried with drum qualities, such as bulk density, solubility, moisture content [37], and particle size, is affected by drum dryers' attributes, such as drying air temperature, rotation speed, feed ability, feed rate to focus, and ambient air quality [37].

3.5 Spray dehydrating

Spray dehydrating is essential to repeatable, continuous, scalable, time-saving, and economical technology for creating dry ultrafine powders [38]. Depending on the nature of the components as well as the required final attributes, numerous types of dryers can be used to dry them [39]. The spray drying procedure maximizes heat transmission and may be utilized for any substance with a liquid-like characteristic [38]. Because of its versatility and speed, it is the most commonly utilized drying technology for various heat-sensitive constituents. Spray drying has an advantage over other dehydration procedures due to its superior product quality, consistent texture, and quick rehydration [39]. Spray-drying technology has also been used in the chemical, pharmaceutical, food, cosmetic, and taste sectors. The approach has several advantages, including being fast, continuous, and repeatable [40]. As a result, it has been effectively implemented at both laboratory and industrial sizes.

3.6 Fluidized-bed drying

By forcing air through the plate's pores, this dryer, which has a drying chamber with a saturated layer design, dries food by heating it until it becomes liquid-like [41]. Typically, crushed materials have a water content of 10–20% in a fluidized bed and 2–5% in the final product [42]. Whey, cocoa, cheese, dessert powders, potato bits, and dried powdered milk are all dried using this drying mechanism [43]. High-water-content foods cannot be dried by fluidized bed drying, but items with low humidity can be dried more gently than other methods, and the materials can be serially ground, chilled, and categorized as they are transported [43].

3.7 Freeze-drying

Although it takes hours, freeze drying has the least amount of protein denaturation as it is done at temperature changes between −30°C and − 40°C and is predicated on vaporization pressure [44]. It also costs between four and six times as much as drying. Freeze drying, correspondingly stated as lyophilization, is a popular method for producing the best quality food solids and powders [45, 46]. Because it functions at lower temperatures and beneath higher pressure, it is the desired technique for dehydrating foods with thermally delicate chemicals and lying to oxidation. Food quality varies as a result of dehydration. Due to the lack of water, oxygen-free atmosphere, and lower temperature conditions, the dehydration of fruits and vegetables through freeze-drying is the optimum method to keep an optimum bio compound content in the final goods [47]. Freeze-drying is a popular method for dehydrating plant-based goods, such as spices, fruits, vegetables, and some unusual meals. Despite its lengthy process time and hefty cost, it is favored for its best final quality [3, 48, 49].

3.8 Novel technology for food dehydration

Consumer desire for processed goods that retain the majority of the unique features of fresh plants has risen in recent years [50]. As a result, drying must be done correctly to keep the plants' flavor, aroma, color, look, and nutritional content as possible [12]. Traditional dehydration processes need extensive drying times and high energy use, resulting in dehydrated items of poor quality [12, 51]. Novel food dehydration methods are a reaction to the newest customer expectations for high-quality dried goods that are also ecologically and economically sustainable [52]. As well as other cutting-edge new drying techniques for food dehydration, the most recent applications of microwave-assisted, solar-assisted, and various drying source-assisted hybrid drying technologies are discussed [53].

3.9 Osmotic dehydration

Osmotic dehydration is a straightforward process that permits fruits to be managed while retaining their natural properties, for example, color, fragrance, nutritional content, and texture [54]. The material derives into contact with a lower water activity mixture during osmotic dehydration when a counter-current mass transfer happens [55] from the product to the solution when water is transported. In the osmotic solution, the soluble solids are assimilated into the foodstuff reversely [55]. As soon as the osmotic pressure of a hypertonic solution is raised, tissue flows water into the solution [55].

3.10 Microwave drying

More and more food items are being dried in the microwave to remove moisture [56]. Microwaves have electromagnetic radiation through frequencies between 300 MHz and 300 GHz and wavelengths between 1 m and 1 mm [57]. Microwave drying provides several benefits over traditional drying processes, including a faster dehydration rate [58]. Microwave drying has numerous downsides, including non-uniform heating, textural degradation, and a restricted microwave penetration depth into materials [59, 60]. A chamber of production in which the material is equipped, a system that measures, controls, and monitors the dehydrating time, and a system that places the product in the dehydrating chamber are the components that make up a typical MWD system [59].

In contrast to drying with hot air, microwaves result in faster drying times and better product quality [61]. When used in place of hot air, microwaves fasten the dehydrating process, reduce oxidation, and increase the dried material's characteristics [62]. These characteristics include density, pore volume, tensile and reconstitution capabilities, color, and the number of bioactive substances present [62]. It is possible that the worth of dried product could be significantly increased with some careful consideration of variables, such as microwave power, an electromagnetic method, temp, humidity ratio, and so on [63].

3.11 Pulsed electric field (PEF)

This method of food preservation is a novel nonthermal method that employs electric current for microbial inactivation [64, 65]. It has been discovered to have little or no negative impact on the quality of food materials [64, 66, 67]. This method primarily processes liquid and semiliquid food products [64, 68, 69]. PEF-assisted drying helps preserve the dried products' physicochemical properties, color, and bioactive compounds (**Figure 2**) [70, 71]. PEF-assisted drying also improves the kinetics of drying and stimulates rehydration. Moreover, also allows selective cell disintegration while keeping the quality of a product [72]. PEF pretreatment inactivates enzymes and microbes; also controls respiratory movement, which may take part in preservation [73]. Despite multiple benefits, the applicability and efficiency of PEF-assisted drying can be enhanced in the future.

3.12 Ball drying

A screw conveyor transports the material that needs to be dried to the top of the drying chamber [74]. It is possible to bypass the conveyor and feed the material directly into the drying chamber; however, doing so will result in the product being fed at a more erratic pace [41]. In addition to that, heated air is constantly being blown into the space that is being examined [75]. The material is brought into contact with heated ceramic balls or other heat-conductive balls while in the dehydrating chamber. The most critical component of the dehydrating process is convection [75]. The large screw located within the chamber spins while the drying procedure is being carried out, and the rate of rotation controls the amount of time the product spends

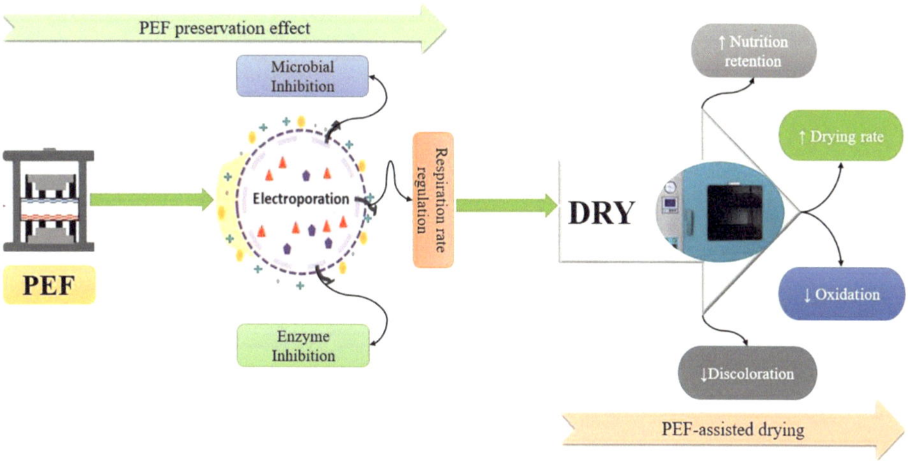

Figure 2.
Proposed mechanism of PEF-assisted drying.

inside the chamber [75]. When the item reaches the end of the chamber, it is taken out along with the balls and then collected [75].

3.13 Ultrasonic drying

The use of ultrasound has shown promise in the inactivation of microorganisms at temperatures close to those of the human body, the enhancement of energy efficiency, the reduction of thermal deterioration of food components, and the maintenance of the site nutritional and sensory integrity of food materials [76–78]. For food dehydration, a technology using stepped-plate transducers to link ultrasonic energy to food samples directly has proven to be an extremely effective method [76, 79]. Plate radiators have a tremendous amount of surface area, which is one of the primary reasons this technology is so advantageous [76, 80]. Other benefits of this technology include the possibility for large-scale industrial applications [76, 81].

3.14 Solar-assisted

Since ancient times, people have relied on the sun's heat to dry out food. Solar energy works by increasing the temperature of the product being dried, which causes an increase in vapor pressure [82]. Vapor pressure is the driving force behind the moisture transfer process [83]. Traditional methods of preserving perishable vegetables include sun dehydration, which involves exposing the vegetables directly to the sun to absorb its radiation [8]. Grapes, prunes, and figs, among other dried fruits, have been preserved using this method to an extensive degree in recent years [8]. Some have been dried using hot-air dehydrators rather than sun drying in recent years because the fruit dries faster with the former, and problems caused by inclement weather are avoided with the latter [84]. Because there is a good chance that future energy limits will be reached, the availability of fossil fuels and natural gas used in dehydration may be severely restricted. If this occurs, a different energy source will be necessary to heat the air used for drying the goods [8]. Heat source from solar energy for hot-air drying appears appealing due to the fact that fruits are typically grown in

Methods	Advantages	Disadvantages
Freeze drying	Polysaccharide does not lose activity	Long processing cycle
Oven drying	Easy to operate	Incomplete drying
Smoking	Added flavors	Difficult to control, slow
Vacuum drying	No oxidation	Low efficiency and high cost
Drum drying	Continuous	It may require modification of the liquid
Spray drying	Fast drying	Easy to decrease the activity of polysaccharides
Ball drying	Rapid, continuous, and comparatively low temperature	Loss of product integrity, difficulty to control
Microwave vacuum drying	Even drying	Low efficiency and low cost
Ultrasound	Reserves antioxidant properties and bioactive compounds	It will affect the antioxidant activity
Pulsed-electric field	• Highly operative for refining drying rates • Short-time handling	Can affect the structure and activity of nutrient
Microwave	• It can improve the yield	Human hazards

Table 1.
Advantages and disadvantages of common drying approaches of food dehydration.

regions that are warm and sunny, as well as the fact that they are dried and harvested through a time in the year when there is an abundance of solar radiation [8].

3.15 Infrared-assisted

The use of hybrid drying methods, such as infrared-freeze drying, has proven to be successful in increasing the efficiency of the dehydration process [85]. The principle behind infrared (IR) dehydration is that IR radiation from a heat source raises the temperature of the object being dried, which helps the moisture evaporate [86]. During the evaporation of water, infrared rays can pierce to a greater depth in the wet sample, which causes the sample's temperature to rise without simultaneously heating the air around it [87]. As the moisture content of the materials decreases and diffuses out into the air, the rate of water diffusion through the material will increase, and the radiation characteristics of the samples will change [88]. Numerous industries frequently use infrared heating because it is widely acknowledged to deliver superior final product quality in addition to countless energy efficiency and cost-effectiveness than convective heating. Several of these industries include [89]. The advantages and disadvantages of common drying techniques are presented in **Table 1**.

4. Hybrid drying technologies

The primary objective in the development of hybrid drying technologies is to slow the deterioration of products while simultaneously producing goods with the desired

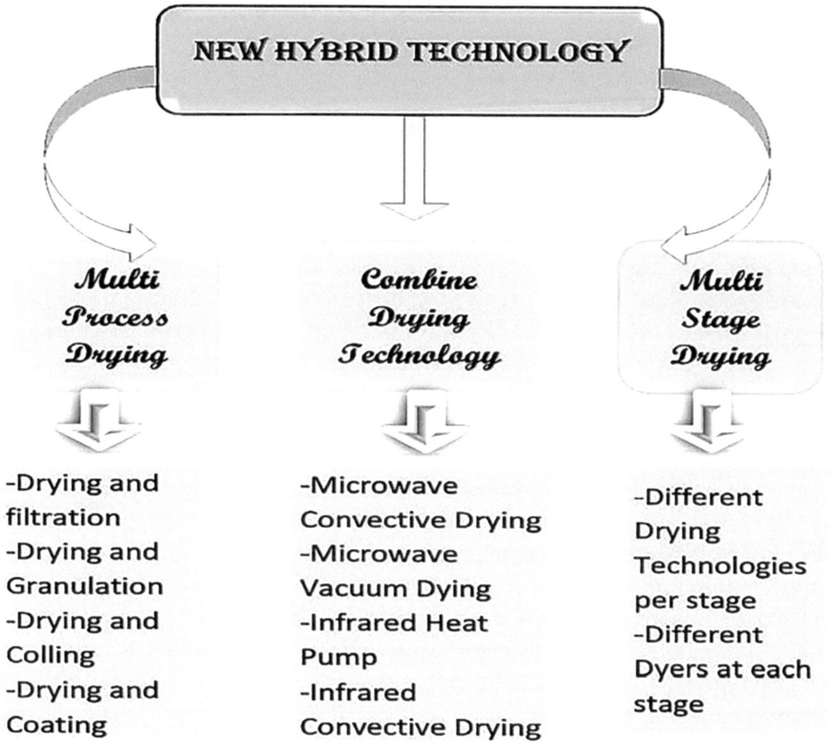

Figure 3.
Hybrid drying technologies scheme of classification.

level of moisture content. When using various drying methods to produce quality-dried goods, some of the most important factors to consider are the characteristics of food quality criteria [90]. **Figure 3** displays an overall Classification of hybrid drying.

The term "hybrid drying technique" refers to the practice of combining two or more distinct drying methods in such a way that they cooperate to reduce the amount of time required and the amount of energy used for drying while maintaining the majority of the product's quality characteristics, such as its flavor, nutrition, color, fragrance, and texture [56]. It has been demonstrated that combined effect and combination drying processes that have been optimized use a low specific amount of energy [91]. The product's characteristics required to be dried play a role in the decision-making process for selecting an appropriate drying method [92]. Efforts are being made to advance hybrid drying technologies to mitigate the disadvantages of more conventional drying procedures, reduce the rate at which products deteriorate, and ensure that end products have the appropriate residual moisture level [56]. The term "hybrid drying" refers to a category of drying processes that include not only those that use multiple modes of heat transmission but also those that use two or more drying phases to achieve the desired level of dryness, product quality, drying time, and production process. A more common definition of hybrid drying is the effective integration or clever combination of two or more conventional drying methods [93]. This one factor may result in the development of an entirely new breed of hybrid drying techniques.

5. Changes in a structure after dehydration of food substance

Various dehydration processes can potentially alter the qualities of foods with a high moisture content [94]. Because moisture removal from a material frequently results in changes to the material properties, these changes play an essential part in the design and prediction of heat and mass transfer processes that occur during dehydration [95]. Additionally, these changes help determine how the methods used to dehydrate products influence the quality properties of dried goods [45]. During the drying process, structural and thermos physical properties such as mass-volume-area-related parameters involved in heat transport are among those that change [96]. Most of these qualities indicate shifts on scales ranging from microscopic to macroscopic in the chemical composition and the structural organization of dried items [97]. Micro and macrostructure observations of the surfaces and cross-sections of fruits and vegetables are frequently included to generate information on the influence of dehydration procedures and circumstances on the textural features of dehydrated samples [56]. These observations can be made at different scales. In addition, morphological parameters are recorded so that an analysis can be performed to determine how the drying processes and environmental factors affect the size and shape of the samples. Drying methods can change primary food attributes: bulk characteristics, flow property, moisture, appearance, aroma, structure, rehydration ability, nutrients, and volatile chemical retention [12]. In terms of nutritional properties, oxygen, extreme heat, and cell injury are common opponents of bioactive component retention during processing. As a result, dehydration can have an effect on the stability of important chemicals in plant-based diets [98]. Phenolic substances may be susceptible to enzymatic breakdown due to polyphenol oxidase activity [98].

The various dehydration techniques unaffected the fundamental structure of polysaccharides [99]. However, different dehydration strategies may change crude polysaccharides' output, protein concentration, and ash concentration [4, 100]. The elimination of vitamin C and carotenoids throughout dehydration procedures is influenced mainly by water concentration and temperature. It is also hypothesized that the structural and thermal physical characteristics of fruits and vegetables change during the drying process based on the chemical composition, the physical organization of the structure, the phase distribution of the system, the internal and exterior pore space represented by the porosity, and other factors [101].

Certain foods subjected to heat treatment have the potential to exhibit an increased total phenolic content [102]. There is a possibility that the drying processes that hasten the breakdown of cellular components are to blame for the rise in the total phenol concentration in samples that have been dried [103–106]. It is possible that the rise can be attributed to the heat-induced breakdown of complex phenolic tannins, which leads to an increased amount of phenolics being extracted [107]. Additionally, the increase in total phenolic content might be explained by the production of Maillard reaction products, which would result in the synthesis of new phenolics from precursors during heat treatments. Compared to conventional dehydration, the color loss that occurs during novel-assisted dehydration is significantly less [108].

6. Future prospective

Osmotic dehydration can extract juice from osmotically concentrated fruits in which juice is ejected from pre-concentrated fruits using osmotic dehydration. It essentially allows the production of high-concentration juice without heat, preserving the nutritional and organoleptic properties of the juice.

7. Conclusion

Dehydration is a simultaneous mass transfer procedure that primarily stimulates the movement of water particles from the meal, resulting in a final item with high sensory characteristics and physiological qualities. In the processing of dried foods, dehydration of food is one of the utmost significant alternative food preservation and treatment techniques. Numerous food dehydration processes are necessary to produce high-quality dried foodstuffs at a low cost. This study discusses recent developments in energy-efficient drying technologies for dehydrating food, including solar, infrared, microwave, and other assisted drying techniques. A new dehydrating method uses less energy and preserves the dried product's chemical, color, taste, flavor, and appearance components. Innovative food drying technology can contribute to environmental protection.

Acknowledgements

We thank Food and Nutrition Society Gilgit Baltistan, Pakistan, for giving free access to journals. The authors also want to acknowledge the support of Guangdong Provincial Key Laboratory of Intelligent Food Manufacturing, Foshan University, Foshan 528225, China (Project ID: 2022B1212010015).

Conflict of interest

There is no conflict of interest.

Author details

Sakhawat Riaz[1], Asifa Kabir[2], Aqsa Haroon[3], Anwar Ali[4,5]
and Muhammad Faisal Manzoor[6,7]*

1 Department of Home Economics, Government College University Faisalabad, Pakistan

2 Department of Nutritional Sciences, Government College University Faisalabad, Pakistan

3 Department of Epidemiology, Universitas Airlangga Surabaya, Indonesia

4 Department of Epidemiology and Health Statistics, Xiangya School of Public Health, Central South University, China

5 Food and Nutrition Society, Gilgit, Baltistan, Pakistan

6 Guangdong Provincial Key Laboratory of Intelligent Food Manufacturing, Foshan University, Foshan, China

7 School of Food Science and Engineering, South China University of Technology, Guangzhou, China

*Address all correspondence to: faisaluos26@gmail.com

References

[1] Mujumdar AS. 23 drying: Principles and practice. Chemical Engineering Handbook. 1667

[2] Gartner E, Maruyama I, Chen J. A new model for the CSH phase formed during the hydration of Portland cements. Cement and Concrete Research. 2017;**97**:95-106

[3] Ahmed N et al. Vegetable proteins: Nutritional value, sustainability, and future perspectives. In: Vegetable Crops-Health Benefits and Cultivation. London, UK, London, UK: IntechOpen; 2021

[4] Ali A, Quratul A, Ayesha S, Waseem K, Munir A, Ahmed B. Bio-Molecular Characteristics of Whey Proteins with Relation to Inflammation. 2021

[5] Kumar D, Kalita P. Reducing postharvest losses during storage of grain crops to strengthen food security in developing countries. Food. 2017;**6**(1):8

[6] Ali A, Riaz S, Sameen A, Naumovski N, Iqbal MW, Rehman A, et al. The disposition of bioactive compounds from fruit waste, their extraction, and analysis using novel technologies: A review. Processes. 2022;**10**(10):2014

[7] Issa IM. Intermediaries' role in urban fresh fruits and vegetables supply chain. In: Salaam DE, editor. Tanzania: A Case Of Kariakoo, Kigamboni And Temeke Stereo Markets. College of Business Education [PHD Thesis]; 2019

[8] Jayaraman K, Gupta DD. Drying of fruits and vegetables. In: Handbook of Industrial Drying. Boca Raton: CRC Press; 2020. pp. 643-690

[9] Tapia MS, Alzamora SM, Chirife J. Effects of water activity (aw) on microbial stability as a hurdle in food preservation. Water activity in foods: Fundamentals and applications. 2020:323-355

[10] Ahmed I, Qazi IM, Jamal S. Developments in osmotic dehydration technique for the preservation of fruits and vegetables. Innovative Food Science & Emerging Technologies. 2016;**34**:29-43

[11] Ali A et al. Relation of electrical stimulation to meat standard. Veterinary Sciences: Research and Reviews. 2021;**7**(1):42-51

[12] Calín-Sánchez Á et al. Comparison of traditional and novel drying techniques and its effect on quality of fruits, vegetables and aromatic herbs. Food. 2020;**9**(9):1261

[13] Pérez-de-Luque A. Interaction of nanomaterials with plants: What do we need for real applications in agriculture? Frontiers in Environmental Science. 2017;**5**:12

[14] Ali A et al. Disaster of housing care amenities versus COVID-19 in urbanized countries: Medication of social science. International Journal. 2021;**7**(5):293

[15] Roobab U, Khan AW, Irfan M, Madni GM, Zeng XA, Nawaz A, et al. Recent developments in ohmic technology for clean label fruit and vegetable processing: An overview. Journal of Food Process Engineering. 2022:e14045

[16] Alp D, Bulantekin Ö. The microbiological quality of various foods dried by applying different drying methods: A review. European Food Research and Technology. 2021;**247**(6):1333-1343

[17] Bialik M et al. Osmotic dehydration and freezing as a suitable pretreatment in

the process of vacuum drying kiwiberry: Drying kinetics and microstructural changes. International Agrophysics. 2020;**34**(2):265-272

[18] Nowak D, Jakubczyk E. The freeze-drying of foods—The characteristic of the process course and the effect of its parameters on the physical properties of food materials. Food. 2020;**9**(10):1488

[19] Kahraman O et al. Drying characteristics and quality attributes of apple slices dried by a non-thermal ultrasonic contact drying method. Ultrasonics Sonochemistry. 2021;**73**:105510

[20] Bhore SJ et al. The avocado (Persea americana mill.): A review and sustainability. Perspectives. 2021

[21] Khalid W et al. Nutrients and bioactive compounds of Sorghum bicolor L. used to prepare functional foods: A review on the efficacy against different chronic disorders. International Journal of Food Properties. 2022;**25**(1):1045-1062

[22] Fortes Da Silva PC. Structure modification of potato slices to reduce oil absorption in deep-fat fried chips (Doctoral dissertation). 2018

[23] Liu G, Jin W. Pervaporation membrane materials: Recent trends and perspectives. Journal of Membrane Science. 2021;**636**:119557

[24] Dosch DE. Synthesis, characterization, scale-up and investigation of structure-property relationships of high energy density materials. Doctoral dissertation, LMU. 2021

[25] Rohmer S, Gerdessen JC, Claassen G. Sustainable supply chain design in the food system with dietary considerations: A multi-objective analysis. European Journal of Operational Research. 2019;**273**(3):1149-1164

[26] Yu F et al. Molybdenum carbide/carbon-based chitosan hydrogel as an effective solar water evaporation accelerator. ACS Sustainable Chemistry & Engineering. 2020;**8**(18):7139-7149

[27] Tudi M et al. Agriculture development, pesticide application and its impact on the environment. International Journal of Environmental Research and Public Health. 2021;**18**(3):1112

[28] Djaeni M et al. Air dehumidification with advance adsorptive materials for food drying: A critical assessment for future prospective. Drying Technology. 2021;**39**(11):1648-1666

[29] Vernekar AK. Desgin of rotary tray system of microwave fruit dehydrate for uniform drying. 2019

[30] Şevik S et al. Mushroom drying with solar assisted heat pump system. Energy Conversion and Management. 2013;**72**:171-178

[31] This H. Note-by-Note Cooking: The Future of Food. Boca Raton: Columbia University Press; 2014

[32] Ali A et al. The burden of cancer, government strategic policies, and challenges in Pakistan: A comprehensive review. Frontiers in Nutrition. 2022;**9**:9

[33] Ongley ED. Control of Water Pollution from Agriculture. Vol. 55. Rome: Food & Agriculture Organization; 1996

[34] Siwal SS et al. Key ingredients and recycling strategy of personal protective equipment (PPE): Towards sustainable solution for the COVID-19 like pandemics. Journal of Environmental Chemical Engineering. 2021;**9**(5):106284

[35] Kumar P et al. Meat snacks: A novel technological perspective. In: Innovations in Traditional Foods. Sawston, Cambridge: Elsevier; 2019. pp. 293-321

[36] Wu H et al. CuMOF-decorated biodegradable nanofibrous membrane: Facile fabrication, high-efficiency filtration/separation and effective antibacterial property. Journal of Industrial and Engineering Chemistry. 2022;**114**:475-482

[37] Pawłowski L. Fundamentals of oxide manufacturing. Industrial Chemistry of Oxides for Emerging Applications. 2018:25

[38] Pardeshi S et al. A meticulous overview on drying-based (spray-, freeze-, and spray-freeze) particle engineering approaches for pharmaceutical technologies. Drying Technology. 2021;**39**(11):1447-1491

[39] Salama AH. Spray drying as an advantageous strategy for enhancing pharmaceuticals bioavailability. Drug Delivery and Translational Research. 2020;**10**(1):1-12

[40] Selvamuthukumaran M. Handbook on Spray Drying Applications for Food Industries. Boca Raton: CRC Press; 2019

[41] Zhang Y, Abatzoglou N. Fundamentals, applications and potentials of ultrasound-assisted drying. Chemical Engineering Research and Design. 2020;**154**:21-46

[42] Eke J, Onwudili JA, Bridgwater AV. Influence of moisture contents on the fast pyrolysis of trommel fines in a bubbling fluidized bed reactor. Waste and Biomass Valorization. 2020;**11**(7):3711-3722

[43] Chuck-Hernandez C, García-Cayuela T, Méndez-Merino E. Dairy-based snacks. In: Snack Foods. Boca Raton: CRC Press; 2022. pp. 417-448

[44] Reubun YTA et al. Freezed drying of Kelor leaves extract (Moringa oleifera lam.). Jurnal Sains dan Kesehatan. 2021;**3**(4):470-474

[45] Bhatta S, Stevanovic Janezic T, Ratti C. Freeze-drying of plant-based foods. Food. 2020;**9**(1):87

[46] Ahmed M et al. Effect of freeze-drying on apple pomace and pomegranate Peel powders used as a source of bioactive ingredients for the development of functional yogurt. Journal of Food Quality. 2022;**2022**:1-9

[47] Bhatkar NS et al. Pre-processed fruits as raw materials: Part I–different forms, process conditions and applications. International Journal of Food Science & Technology. 2022;**57**(8):4945-4962

[48] Oyinloye TM, Yoon WB. Effect of freeze-drying on quality and grinding process of food produce: A review. PRO. 2020;**8**(3):354

[49] Babar Q et al. Novel treatment strategy against COVID-19 through anti-inflammatory, antioxidant and Immunostimulatory properties of the B vitamin complex. In: B-Complex Vitamins-Sources, Intakes and Novel Applications. London, UK, London, UK: Intechopen; 2021

[50] Birch CS, Bonwick GA. Ensuring the future of functional foods. International Journal of Food Science & Technology. 2019;**54**(5):1467-1485

[51] Ali A et al. Novel therapeutic drug strategies to tackle immune-oncological challenges faced by cancer patients during COVID-19. Expert Review of Anticancer Therapy. 2021;**21**(12):1371-1383

[52] Grahl S et al. Consumer-oriented product development: The conceptualization of novel food products based on spirulina (Arthrospira platensis) and resulting consumer expectations. Journal of Food Quality. 2018;**2018**:1-11

[53] Juliano P, Reyes-De-Corcuera JI. Food engineering innovations across the food supply chain: Debrief and learnings from the ICEF13 congress and the future of food engineering. In: Food Engineering Innovations across the Food Supply Chain. Cambridge, Massachusetts: Elsevier; 2022. pp. 431-476

[54] Pichardo-Romero D et al. Current advances in biofouling mitigation in membranes for water treatment: An overview. PRO. 2020;**8**(2):182

[55] Lewicki PP, Lenart A. Osmotic dehydration of fruits and vegetables. In: Handbook of Industrial Drying. Boca Raton: CRC press; 2020. pp. 691-713

[56] Zielinska M et al. Review of recent applications and research progress in hybrid and combined microwave-assisted drying of food products: Quality properties. Critical Reviews in Food Science and Nutrition. 2020;**60**(13):2212-2264

[57] Mirbeik-Sabzevari A, Tavassolian N. Tumor detection using millimeter-wave technology: Differentiating between benign lesions and cancer tissues. IEEE Microwave Magazine. 2019;**20**(8):30-43

[58] Pongpichaiudom A, Songsermpong S. Characterization of frying, microwave-drying, infrared-drying, and hot-air drying on protein-enriched, instant noodle microstructure, and qualities. Journal of Food Processing and Preservation. 2018;**42**(3):e13560

[59] Laguerre JC, Hamoud-Agha MM. Microwave food processing: Principles and applications. Thermal Food Engineering Operations. 2022;**11**:301-347

[60] Ahmad N et al. Impact of thermal extrusion and microwave vacuum drying on fatty acids profile during fish powder preparation. Food Science & Nutrition. 2021;**9**(5):2743-2753

[61] Jia Y et al. Influence of three different drying techniques on persimmon chips' characteristics: A comparison study among hot-air, combined hot-air-microwave, and vacuum-freeze drying techniques. Food and Bioproducts Processing. 2019;**118**:67-76

[62] Zeng Y et al. Effects of far-infrared radiation temperature on drying characteristics, water status, microstructure and quality of kiwifruit slices. Journal of Food Measurement and Characterization. 2019;**13**(4):3086-3096

[63] Khan MIH et al. Modelling of simultaneous heat and mass transfer considering the spatial distribution of air velocity during intermittent microwave convective drying. International Journal of Heat and Mass Transfer. 2020;**153**:119668

[64] Arshad RN et al. Electrical systems for pulsed electric field applications in the food industry: An engineering perspective. Trends in Food Science & Technology. 2020;**104**:1-13

[65] Manzoor MF et al. Impact of pulsed electric field on rheological, structural, and physicochemical properties of almond milk. Journal of Food Process Engineering. 2019;**42**:e13299

[66] Manzoor MF et al. Effect of pulsed electric field and thermal treatments on the bioactive compounds, enzymes, microbial, and physical stability of almond milk during storage. Journal of Food Processing Preservation. 2020;**44**(7):e14541

[67] Ahmed Z et al. Impact of pulsed electric field treatments on the growth parameters of wheat seeds and nutritional properties of their wheat plantlets juice. Food Science Nutrition. 2020;**8**(5):2490-2500

[68] Manzoor MF et al. Combined impact of pulsed electric field and ultrasound on bioactive compounds

and FT-IR analysis of almond extract. Journal of Food Science and Technology. 2019;**56**(5):2355-2364

[69] Ahmed Z et al. Study the impact of ultra-sonication and pulsed electric field on the quality of wheat plantlet juice through FTIR and SERS. Ultrasonics Sonochemistry. 2021;**76**:105648

[70] Rahaman A et al. Impact of pulsed electric field treatment on drying kinetics, mass transfer, colour parameters and microstructure of plum. Journal of Food Science and Technology. 2019;**56**(5):2670-2678

[71] Rahaman A et al. Effect of pulsed electric fields processing on physiochemical properties and bioactive compounds of apricot juice. Journal of Food Process Engineering. 2020;**43**(8):e13449

[72] Rahaman A et al. Combined effect of pulsed electric fields and ultrasound on mass energy transfer and diffusion coefficient of plum. Heat and Mass Transfer. 2021;**57**:1-9

[73] Manzoor MF et al. Novel processing techniques and spinach juice: Quality and safety improvements. Journal of Food Science. 2020;**85**(4):1018-1026

[74] Zhang Y et al. Dynamics of heat-sensitive pharmaceutical granules dried in a horizontal fluidized bed combined with a screw conveyor. Chemical Engineering and Processing-Process Intensification. 2021;**167**:108516

[75] Rahman MS, Perera CO. Drying methods used in food preservation. In: Handbook of Food Preservation. Boca Raton: CRC Press; 2020. pp. 427-442

[76] Bhargava N et al. Advances in application of ultrasound in food processing: A review. Ultrasonics Sonochemistry. 2021;**70**:105293

[77] Rahaman A et al. Ultrasound based modification and structural-functional analysis of corn and cassava starch. Ultrasonics Sonochemistry. 2021;**80**:105795

[78] Rahaman A et al. Influence of ultrasound-assisted osmotic dehydration on texture, bioactive compounds and metabolites analysis of plum. Ultrasonics Sonochemistry. 2019;**58**:104643

[79] Manzoor MF et al. Thermosonication effect on bioactive compounds, enzymes activity, particle size, microbial load, and sensory properties of almond (Prunus dulcis) milk. Ultrasonics Sonochemistry. 2021;**78**:105705

[80] Manzoor MF et al. Impact of high-intensity thermosonication treatment on spinach juice: Bioactive compounds, rheological, microbial, and enzymatic activities. Ultrasonics Sonochemistry. 2021;**78**:105740

[81] Manzoor MF et al. Effect of dielectric barrier discharge plasma, ultra-sonication, and thermal processing on the rheological and functional properties of sugarcane juice. Journal of Food Science. 2020;**85**(11):3823-3832

[82] Nain S. Evolution and advancements in solar drying technologies: A review. Latest Trends in Renewable Energy Technologies. 2021:249-260

[83] Xu C, Li S, Zou K. Study of heat and moisture transfer in internal and external wall insulation configurations. Journal of Building Engineering. 2019;**24**:100724

[84] Patidar A, Vishwakarma S, Meena D. Traditional and recent development of pretreatment and drying process of grapes during raisin production: A review of novel pretreatment and drying methods of grapes. Food Frontiers. 2021;**2**(1):46-61

[85] Liu W et al. A novel strategy for improving drying efficiency and quality of cream mushroom soup based on microwave pre-gelatinization and infrared freeze-drying. Innovative Food Science & Emerging Technologies. 2020;**66**:102516

[86] Athira V et al. Advances in drying techniques for retention of antioxidants in agro produces. Critical Reviews in Food Science and Nutrition. 2022:1-17

[87] Chang A et al. Short-and medium-wave infrared drying of cantaloupe (Cucumis melon L.) slices: Drying kinetics and process parameter optimization. PRO. 2022;**10**(1):114

[88] Wen Y-X et al. Effect of infrared radiation-hot air (IR-HA) drying on kinetics and quality changes of star anise (Illicium verum). Drying Technology. 2021;**39**(1):90-103

[89] Al-Waeli AH et al. Photovoltaic/Thermal (PV/T) Systems: Principles, Design, and Applications. Switzerland: Springer Nature; 2019

[90] Acar C, Dincer I, Mujumdar A. A comprehensive review of recent advances in renewable-based drying technologies for a sustainable future. Drying Technology. 2022;**40**(6):1029-1050

[91] Qaidi SM et al. Rubberized geopolymer composites: A comprehensive review. Ceramics International. 2022;**48**:24234-24259

[92] Przybył K, Gawałek J, Koszela K. Application of artificial neural network for the quality-based classification of spray-dried rhubarb juice powders. Journal of Food Science and Technology. 2020:1-11

[93] Lamidi RO et al. Recent advances in sustainable drying of agricultural produce: A review. Applied Energy. 2019;**233**:367-385

[94] Xu Y et al. A comparative evaluation of nutritional properties, antioxidant capacity and physical characteristics of cabbage (Brassica oleracea var. capitate var L.) subjected to different drying methods. Food Chemistry. 2020;**309**:124935

[95] Pecha MB et al. Progress in understanding the four dominant intra-particle phenomena of lignocellulose pyrolysis: Chemical reactions, heat transfer, mass transfer, and phase change. Green Chemistry. 2019;**21**(11):2868-2898

[96] Aydogan OC. Changes in Thermal Properties during the Growing Season to Predict the Apple Harvest Time and Monitor the Quality. USA: The Pennsylvania State University; 2018

[97] Ghebremedhin M, Seiffert S, Vilgis TA. Molecular behavior of fluid gels–the crucial role of edges and particle surface in macroscopic properties. Food & Function. 2022;**13**(13):6902-6922

[98] Schumann M, Holm J, Brinker A. Effects of feeding an all-plant diet on rainbow trout performance and solid waste characteristics. Aquaculture Nutrition. 2022;**2022**:1-11

[99] Yurtsever A et al. Probing the structural details of chitin nanocrystal–water interfaces by three-dimensional atomic force microscopy. Small Methods. 2022;**6**:2200320

[100] Li Y et al. Morphological and structural changes in thermally-induced soybean protein isolate xerogels modulated by soybean polysaccharide concentration. Food Hydrocolloids. 2022;**133**:107967

[101] Gomez-Lopez VM et al. Guidelines on reporting treatment conditions for emerging technologies in food processing. Critical Reviews in Food Science and Nutrition. 2022;**62**(21):5925-5949

[102] Dibanda RF, Akdowa EP, Tongwa QM. Effect of microwave blanching on antioxidant activity, phenolic compounds and browning behaviour of some fruit peelings. Food Chemistry. 2020;**302**:125308

[103] Barbhuiya RI, Singha P, Singh SK. A comprehensive review on impact of non-thermal processing on the structural changes of food components. Food Research International. 2021;**149**:110647

[104] Manzoor MF, Hussain A, Naumovski N, Ranjha MMAN, Ahmad N, Karrar E, et al. A narrative review of recent advances in rapid assessment of anthocyanins in agricultural and food products. Frontiers in Nutrition. 2022;**9**

[105] Manzoor MF, Hussain A, Tazeddinova D, Abylgazinova A, Xu B. Assessing the nutritional-value-based therapeutic potentials and non-destructive approaches for mulberry fruit assessment: An overview. Computational Intelligence and Neuroscience. 2022;**2022**

[106] Manzoor MF, Hussain A, Sameen A, Sahar A, Khan S, Siddique R, et al. Novel extraction, rapid assessment and bioavailability improvement of quercetin: A review. Ultrasonics Sonochemistry. 2021;**78**:105686

[107] Kataria A, Sharma S, Dar B. Changes in phenolic compounds, antioxidant potential and antinutritional factors of Teff (Eragrostis tef) during different thermal processing methods. International Journal of Food Science & Technology. 2021;**57**:6893-6902

[108] Cui H et al. Formation and fate of Amadori rearrangement products in Maillard reaction. Trends in Food Science & Technology. 2021;**115**:391-408

Chapter 3

A Review of Drying Methods Assisted by Infrared Radiation, Microwave and Radio Frequency

Nguyen Hay, Le Quang Huy and Pham Van Kien

Abstract

The study focused on reviewing modern and effective drying methods assisted by infrared radiation, microwave and radio frequency. In which, the drying results of previous studies were reviewed to clarify the drying efficiency of drying methods with the support of infrared radiation, microwave and radio frequency. The review results showed that the radiant heating mechanism of infrared radiation and the volumetric heating mechanism of microwave and radio frequency supported the process of material heating and moisture diffusion within the material. As a result, the drying process achieved high drying efficiency, the drying time was significantly shortened and the quality of the dried products was improved both in terms of sensory quality and nutritional quality. The study of the application of infrared radiation, microwave and radio frequency in drying technique had a high scientific, technological and practical significance. This would be the foundation for finding suitable drying methods and drying modes to improve drying efficiency as well as the quality of dried products.

Keywords: infrared radiation, microwave, radio frequency, drying efficiency, volumetric heating mechanism, product quality, energy consumption

1. Introduction

The fresh harvested agricultural products often contain high moisture content, which can easily spoil the products by the risk of decay as oxidation, hydrolyzation, and microbial spoilage during processing, transit, and storage [1]. Nowadays prolonging the storage time of products and increasing the product quality both in terms of sensory quality and nutritional quality has been becoming an important target of a sustainable agriculture.

Drying has played an important role in food processing and preservation. Drying can reduce the water activity in order to improve the stability of drying products and long-time preservation. Because the harmful microbial activity would be significantly decreased at lowering water activity during storage [2]. In the drying process, the drying methods and drying parameters should be considered to achieve the effective drying process as drying rate and product quality [3].

Generally, a suitable and effective drying technique should achieve several requirements such as high drying rate, required final moisture content, high retention in color, appearance, flavor, long shelf life and high ingredient retention of dried products, as well as low energy consumption [4]. In the drying of agricultural products, hot air drying, solar drying, vacuum drying, freeze drying and fluidized bed drying are popular conventional drying techniques [5, 6]. In the conventional drying, the heat from drying air will transfer to the drying material and the moisture of the material evaporates back to the drying air by convection and conduction. However, the convective drying method has many drawbacks such as longer drying time, low product quality, low efficiency and high energy consumption [7, 8]. Hot air and solar drying with the heating convection mechanism have been becoming common drying methods because of low cost of equipment, large drying capacity and simple operation [9]. However, the hot air drying offers a time and energy consuming because of the large amount of heat losses to the air [10]. The solar drying can save energy consumption, but the solar dryer works in natural environment, which makes dried products be affected by insects and microorganisms. Besides, hot air drying and solar drying methods can easily cause shrinkage, surface hardening, microstructure collapse, high color change index and nutrient degradation [11]. Vacuum and freeze-drying methods with low drying temperature are suitable drying methods for the preservation of nutritional components, but the requirement of special operation condition as low temperatures or pressures cause high energy consumption and cost of equipment. Fluidized bed drying could improve the energy efficiency with the contact between the air and materials, but this drying method is almost appropriate for granular food [5]. Thus, many new drying techniques have been created by the combination between the conventional drying and infrared radiation (IR) drying, microwave (MW) drying, radio frequency (RF) drying and ultrasound wave drying [12]. In which, effective heating mechanism of IR, MW and RF would support the drying process effectively and could enhance the efficiency of drying process as high drying rate, high dried product quality and low energy consumption. However, the application of these hybrid drying methods has been studied and continues to be studied to perfect both drying technology and drying parameters to achieve the highest drying efficiency.

2. The heating principle of infrared radiation, microwave and radio frequency

2.1 The heating principle of infrared radiation

The infrared region lies between the visible and microwave region of the electromagnetic spectrum. Infrared waves have a wavelength ranging from 0.75 to 1000 μm. Generally, IR can be divided into three different categories such as: near-infrared (NIR), mid-infrared (MIR) and far-infrared (FIR) corresponding to the wavelength of 0.75–2 μm, 2–4 μm and 4–1000 μm (see **Figure 1**) [13].

The effect of the electromagnetic radiation on drying material surface causes changes in the electronic, rotational and vibrational states of atoms and molecules [14]. The different type of electromagnetic radiation wavelength has different heating mechanism. The heating mechanism of IR radiation is the changes in the vibrational state of the molecule in material. When the materials absorb IR radiation energy, the constituent molecules will vibrate at a frequency of 60,000–150,000 MHz and the intermolecular friction occurs which results in the internal heating within material [15]. In which, the surface temperature increases quickly, which is much faster than

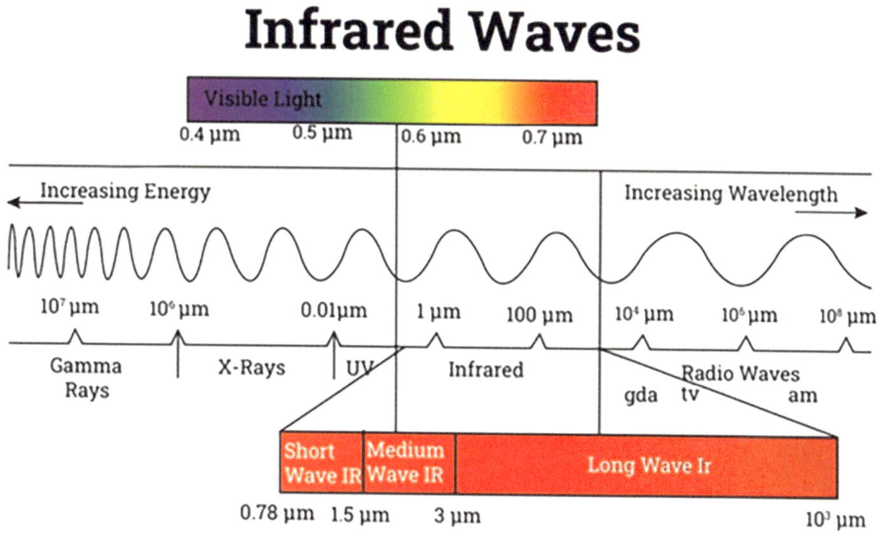

Figure 1.
The infrared region.

that in conduction and convection heat transfer [16]. IR radiation can propagate into the material without any medium and heating the surrounding air [17]. The most important advantages of infrared radiation are preventing the energy losses and maintaining the quality of the drying product.

The IR radiation has a specific penetration depth (PD) at which the radiation intensity inside the material reaches the value of approximately 37% of its value at the surface [18, 19]. PD of IR radiation depends on the composition of material such as solid, liquid or gas, and the properties of material such as density, porosity and water content [20].

2.2 The heating principle of microwave and radio frequency

Radio frequency (RF) and microwave (MW) are two kinds of electromagnetic waves. In which, RF has frequencies varying from 10 to 300 MHz and only three radio frequencies of 13.56 MHz ± 6.68 kHz, 27.12 MHz ±160.00 kHz and 40.68 MHz ± 20.00 kHz were applied in industrial, scientific, and medical applications [21]. Microwave has the frequency range of 300 MHz to 300 GHz and MW with frequency of 2450 MHz was the most commonly used in the heating technique. The frequency range of RF and MW was described in **Figure 2**.

RF heating mechanism is volumetric heating and is described in **Figure 3**. In which, the RF is generated between two electrodes of an RF operator. The material is placed between the electrodes and absorbed RF energy. RF energy causes the wet molecules within material to migrate and rotate continuously at high speeds under the effect of the RF alternating electric field. The volumetric heat was generated within the material by the friction resulting from the movement of the molecules and the space-charge displacement.

In the case of microwaves, the electric field interacts with molecules of material via two modes as dipolar rotation and ionic conduction (**Figure 4**). In dipolar rotation, the molecules within material rotate back and forth constantly under the

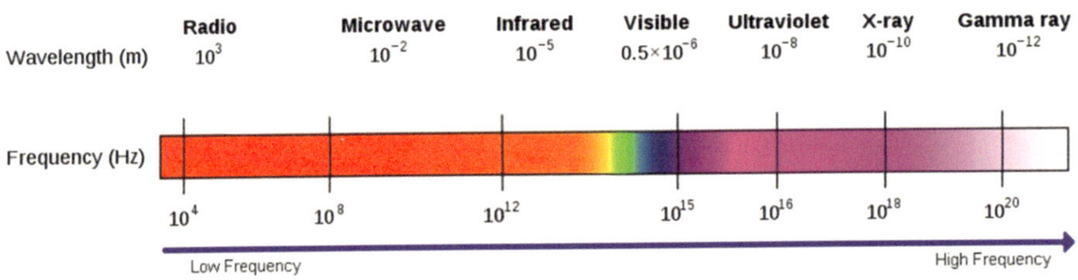

Figure 2.
The frequency range of electromagnetic wave.

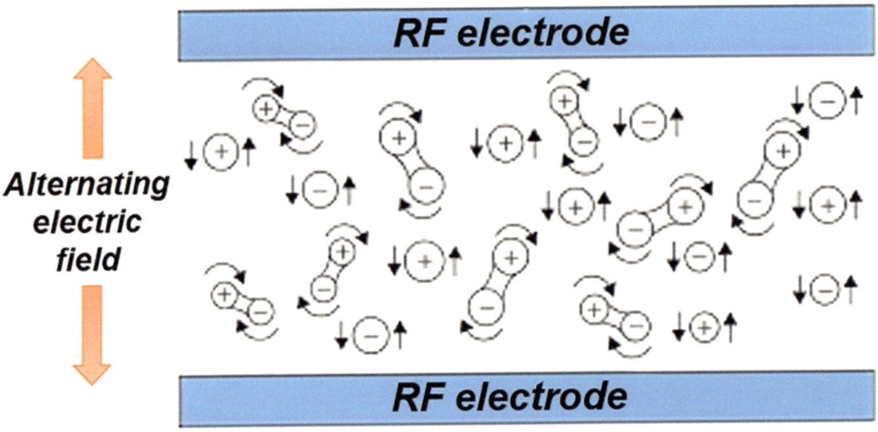

Figure 3.
RF heating mechanism.

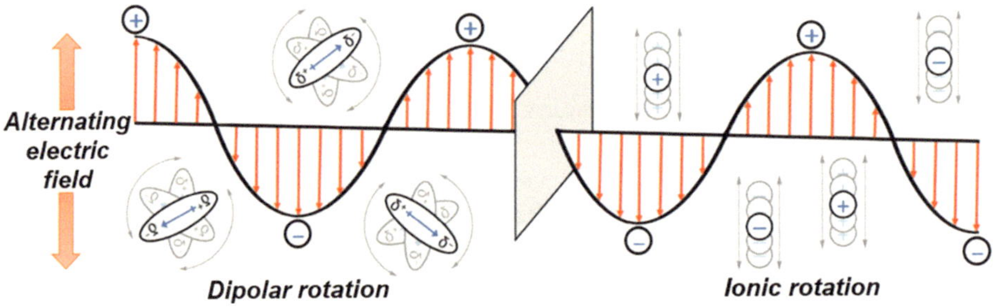

Figure 4.
MW heating mechanism.

ever-oscillating electric field. The friction of rotating molecules results in heat generation through the material mass. In ionic conduction, a free ion or ionic species moves translationally through space under the alternating electric field. The friction of these moving species results in heat generation.

Both RF and MW heating mechanism is volumetric heating. In which, the heating rate is high and the moisture gradient and temperature gradient are in the same direction, which supports the moisture diffusion. Thus, the drying rate and the quality of the drying product would significantly increase.

3. Review of the drying techniques using infrared radiation, microwave and radio frequency

3.1 Review of the drying techniques using infrared radiation

The drying techniques using infrared radiation has several advantages compared to conventional drying system such as higher heating rate, shorter drying time and higher drying product quality [22]. Thus, the drying techniques using infrared radiation have become one of the popular methods of drying of food, vegetables, grains, fruits and other high-value products [23–25]. Combining IR with other drying methods as hot air drying, heat pump drying, vacuum drying and freeze drying can give an effective drying process [26–29].

Wanyo et al. [30] conducted the experimental drying of mulberry leaves by two drying methods as far IR-hot air drying and hot air-only drying. The results showed that the far IR-hot air-drying time was only 50 min and the higher total phenolic content within mulberry leaves was retained, while hot air-drying time required 6 hours. Chen et al. [31] studied the drying kinetics and quality product in drying of jujube slices by hot-air and short-and medium-wave infrared radiation and the results showed that the short- and medium-wave IR drying time could reduce 17–67% as compared with hot air drying. Orikasa et al. [32] reported that the far infrared (FIR) drying of Komatsuna leaves consumed 17% less energy consumption and achieved the higher product quality with higher cyclic adenosine monophosphate content within Komatsuna leaves after drying as compared with hot air drying.

The combination between vacuum drying and IR heating could result in high drying rate, improvement of energy efficiency and product quality [33]. Salehi and Kashaninejad [34] studied the drying kinetics during the combined infrared-vacuum drying of lemon slices and the results confirmed that the moisture diffusivity within material increased from 2.92×10^{-10} to 1.58×10^{-9} m^2/s and the color change index decreased appreciably when the IR lamp power was increased from 300 to 400 W, which improved the drying rate considerably Salehi and Kashaninejad [35] investigated the effects of IR-vacuum drying parameters with infrared power of 300–400 W on the drying kinetics of grapefruit slices. The results were that the effective moisture diffusivity increased in the range of 5.83×10^{-10} and 2.13×10^{-9} m^2/s and the color change intensity of dried material decreased as IR power increased.

The combined IR-freeze drying could improve the freeze-drying process that could reduce drying time and energy consumption. Wu et al. [28] compared the effect of the freeze drying and the IR-freeze drying of Cordyceps militaries, in which, the IR-freeze dryer used infrared lamps replacing the electric heating plate and the drying parameters as drying time, energy consumption and nutritional properties were considered. The results indicated that IR-freeze drying achieved high-quality dried products and more effective drying process such as the drying time could be reduced 7.21–17.78% and the energy consumption was reduced 1.88–18.37% at a constant drying temperature in comparison with the freeze drying.

Khampakool et al. [36] conducted the producing process of banana snacks using IR-assisted freeze drying, in which, the drying kinetics, energy consumption, and product quality were estimated. The results indicated that the continuous IR-assisted freeze drying could save the electrical energy consumption appreciably and reduce the drying time up to 213 min as compared with freeze drying (696 min). Besides, the crispness of banana snacks could be improved in IR-assisted freeze drying.

Antal (2015) [37] compared the effect of three drying of apple cubes methods as IR-assisted freeze drying, hot air-assisted freeze drying and single-stage freeze drying and the results reported that the drying time was reduced by 45.5 and 27.3% and energy consumption by 45.1 and 34.5% corresponding to IR-assisted freeze drying, hot air–assisted freeze drying as compared with FD.

Krsti (2002) [38] proposed a prototype rotary dryer with combining IR with heat pump drying method for drying of herbs and vegetables (leaves of birch, rosebay willowherb, dandelion, red beet and carrot). The results showed that the drying time was shortened significantly and the quality of the dried products as color and ingredient retention within the material got requirement.

Based on the results of the previous studies mentioned above, IR heating is an effective supporting technique when combined with other drying methods such as hot air drying, heat pump drying, vacuum drying and freeze drying to significantly improve drying efficiency including drying rate and product quality as well as saving energy consumption of drying process as compared with drying methods without IR assistance. In particular, IR heating assistance could reduce the drying time by 67% and the energy consumption by 17% as compared to hot-air drying. The combination between vacuum drying and IR heating could significantly improve the effective moisture diffusivity of drying material and the quality of the dried products. In the combined IR-freeze drying, the drying time could be reduced by 45.5% and the energy consumption reduced by 45.1% in comparison with the freeze drying. The combining IR with heat pump drying method could shorten the drying time significantly and retain the color and ingredient content within the material. Besides, the assistance of IR improved the quality of the dried products as color and nutrients, that is one of the most important factors of drying technique.

3.2 Review of the drying techniques using microwave

The volumetric heating mechanism of MW has many advantages compared to conventional methods. MW has been widely applied in drying technique for drying food, vegetable, fruit and high-value material. When MW combines with other conventional drying methods, MW heating mechanism can improve the efficiency of drying process as heating rate, drying rate and quality of dried products [39]. There have been many studies of drying techniques using MW to find out the suitable dying methods and drying parameters to improve the drying process and quality of dried products.

Soysal (2004) [40] studied the effects of MW output power on drying time, drying rate and color of product in MW drying of parsley leaves, in which, seven different MW output powers were used ranging from 360 to 900 W. The results indicated that drying time considerably decreased with the increase in the MW output power and a good green color was maintained close to that of the original fresh parsley leaves.

Yanyang et al. [41] studied the wild cabbage drying method by a combination of hot-air drying and microwave vacuum drying. The results confirmed that the combination of drying with hot air drying followed by microwave-vacuum drying could shorten drying time and greatly retain the content of chlorophyll and ascorbic acid within the dried product.

Therdthai and Zhou (2009) [42] carried out the drying of mint leaves with MW-vacuum drying (8.0 W/g, 9.6 W/g and 11.2 W/g at pressure 13.33 kPa) and hot air drying (60°C and 70°C). The MW-vacuum drying could reduce drying time by

85–90% as compared with the hot air drying. The color of dried mint leaves was retained in light green/yellow color while the hot air-dried mint leaves were changed in to dark brown.

Wang et al., (2009) [43] dehydrated instant vegetable soup mix using the MW–freeze drying. The vegetable soup was successfully dried and the drying process achieved a shorter drying time as MW power increased. The energy consumption of MW-freeze dryer reduced significantly as compared to freeze dryer.

Patil et al., (2015) [44] carried out the experimental drying of green leafy vegetable (fenugreek, coriander, spinach, mint, shepu and curry leaves). The results reported that as the MW power increased from 135 to 675 W, the drying time reduced appreciably by 64% and the dried green leafy vegetables could be stored for about 21 days in metalized polyester under the condition of 45°C and 95% RH. Besides, the shelf life of dried green leafy vegetables could last up to 6 months as being stored in metalized polyester under the condition of 65% RH and 30°C.

Akal and Kahveci (2016) [45] conducted the microwave drying of carrot slices, in which, MW power levels were 350, 460 and 600 W and thickness of carrot slices was 1 and 2 cm. The results showed that the drying rate increased as the drying thickness decreased and MW power increased. When MW power increased from 350 to 600 W, the drying time reduced up to 50%.

Horuz et al., (2017) [46] determined the effect of hybrid (MW convectional) and convectional drying of sour cherries, in which, the MW power was 120, 150 and 180 W and hot air was 50, 60 and 70°C. The results determined that the energy consumption efficiency of hybrid drying technique was higher than the convectional drying method and in hybrid drying, the drying time was reduced significantly and the dried products could achieve the higher quality parameters as total phenoliccontent, antioxidant capacity and vitamin C.

Deepika and Sutar (2018) [47] conducted the drying of lemon slices using IR, MW and hot air combination, in which, IR–hot air drying was implemented and followed by MW–hot air to complete the drying process. The results found that the drying process could save energy consumption and drying time and the quality of the dried product was also maintained as compared to hot air drying.

Rodriguez et al., (2019) [48] evaluated the effect of solar drying and microwave drying of raspberries. The results showed that MW drying method could significantly reduce the drying time as compared to solar drying. The quality of dried products indicated that in both drying methods, the surface color of dried products was retained close to the color of fresh raspberries.

Jinwoo et al., (2021) [49] studied the influence of microwave power in MW-assisted freeze drying of biopharmaceuticals. The results showed that the combination between MW volumetric heating and heat conduction reduced drying times by 80% as compared to freeze drying. While the quality of dried products in MW-assisted freeze drying was close to freeze drying.

Based on the results of the previous studies mentioned above, the drying technique using MW showed its outstanding advantages that have greatly improved drying efficiencies such as appreciably shortening drying time, saving energy consumption and improving the quality of dried products in terms of color and the content of important nutrients within the dried products. Drying technique using MW could be considered as a suitable and effective drying method for many types of materials such as food, vegetable, fruit and high-value material. In particular, MW heating assistance could reduce the drying time by 64% as increasing MW power in specific value range. The combining MW with other drying methods as hot-air drying, vacuum drying, solar drying and freeze drying could improve the drying rate by shortening the drying time by 80% and achieve the higher quality dried products as color, vitamin and nutrients as compared to the drying method without MW assistance.

3.3 Review of the drying techniques using radio frequency

Radio frequency has been studied and applied for various purposes in food processing such as cooking [50], tempering [51], stabilization [52], disinfestation [53], pasteurization [54], roasting [55] and drying [56]. Drying using RF was one of the most important and effective application of RF. RF has been applied in combining with other drying methods as hot air drying, heat pump drying, vacuum drying and freeze drying in order to achieve the high drying efficiency.

Roknul et al. (2014) [57] compared the effect of four drying methods including hot-air drying, hot-air-assisted IR drying, hot-air-assisted MW drying and hot-air-assisted RF drying of stem lettuce slices, in which, the drying characteristics and quality of stem lettuce slices were considered. The results showed that the drying time of hot-air-assisted RF drying was the shortest (120 min), following by hot-air-assisted MW drying (140 min), hot-air-assisted IR drying (180 min), and hot-air drying (360 min). In the hot-air-assisted RF drying, the heating was uniform and the color change index of dried samples got the smallest value.

Zhou et al., (2018) [58] carried out the experimental drying of kiwifruit slices using RF-vacuum drying and hot air drying, in which, RF had the working parameter of 27.12 MHz, 3 kW. The results demonstrated that RF-vacuum drying was supported by RF volumetric heating and achieved a rapid drying resulting in 65% reduction of hot air drying (60°C) time. Moreover, RF-vacuum drying dried kiwifruit slices retained a better color stability, higher content of vitamin C as compared with hot-air-dried samples.

Zhou et al., (2018) [59] compared three drying methods including hot air drying, vacuum drying and hot air-assisted RF drying, which were experimentally compared and analyzed for drying of in-shell walnuts. The results showed that the drying time of the hot air-assisted RF drying method was the shortest (138 min), followed by vacuum drying method (185 min) and hot air-drying method (300 min). The walnuts after hot air-assisted RF drying process contained more unsaturated fatty acid than those by hot air-drying process and the vacuum drying, and hot air-assisted RF drying had little effect on the total antioxidant capacity and total phenolic concentration of walnuts during the drying process and storage.

Zhou et al., (2019) [60] determined the drying time, energy efficiency and product quality of three drying methods as RF-vacuum drying, hot air drying and RF-vacuum + hot air drying in drying of 6 mm thick kiwifruit slices. The results indicated that the total drying time of RF-vacuum drying was the shortest (480 min), followed by RF-vacuum + hot air drying (600 min; 20% longer) and hot air drying (900 min, almost double). In the RF-vacuum + hot air-drying process, a more uniform moisture distribution within the fruit slices and better product quality as color retention, shrinkage ratio was achieved. Besides, RF-vacuum + hot air-drying method could save the average energy efficiency up to 22.93% as compared to hot air-drying method.

Zhang et al., (2019) [61] studied the hot air-assisted RF as the second stage drying method for mango slices, in which, hot-air drying was used in the first drying stage to reduce the moisture content to about 40% (w.b), then hot air-assisted RF drying was used to reduce moisture content to 18%. The results showed that the drying time of the two-stage drying process was about 5 hours, which was lower than that of hot-air drying (8 hr) or vacuum drying (7 hr). The quality of dried mango slices after the two-stage drying was better than that after hot-air drying, and close to that after vacuum drying, in which, the minor vitamin C retention rate achieved 91%.

Ran et al., (2019) [62] applied the hybrid drying method of RF-assisted vacuum drying to produce chicken powders. The results showed that the total drying time of RF-assisted vacuum drying was the shorter (100 min) while that for vacuum drying was 180 min. Besides, RF-assisted vacuum-dried chicken powders had a high quality with a maximum umami flavor among the obtained powders.

Peng et al., (2019) [63] proposed a hybrid drying method for apple slices (6 mm thickness), in which, air jet impingement (65°C) was applied at the first drying stage, and hot air-assisted RF (hot air temperature of 60°C) was used at the second drying stage to complete the drying process. The result found that apple slices dried by the air jet impingement combined hot air-assisted RF drying achieved the shortest drying time and the best product quality including color, total phenolic and Vc retention.

Wang et al., (2020) [64] conducted the experimental drying of inshell hazelnuts with hot air drying and hot air-assisted RF drying. The results found that as compared to hot air drying, hot air-assisted RF drying achieved much higher drying rate and effective moisture diffusion with lower energy consumption. Besides, hot air-assisted RF-dried nuts retained higher total phenolic content (0.43 mg GAE/g dry kernel) than that of hot air drying.

Wang et al., (2021) [65] investigated the efficiency of hot air-assisted RF drying of carrot. In which, the hot air (60°C)-assisted RF heating at the electrode gap of 100 mm was used for drying of carrot slices for 270 min first and then followed by hot air drying to complete the drying process. The results showed that the combined drying method could improve the drying rate up to 30% and maintain higher content of heat-sensitive ingredient within carrot slices.

Shewale et al., (2021) [66] studied the efficiency of the sequential drying method using RF and low-humidity air (LHA, 40°C) for drying of apple slices. The combined drying method of LHA + RF could reduce the drying time by 37% and energy consumption by 52% as compared to LHA. The LHA + RF dried apple slices were preserved the polyphenols (98%), flavonoids (87%) and ascorbic acid (77%) and the color change index was so small ($\Delta E = 7.4 \pm 0.7$).

Le Anh Duc et al., (2022) [67] studied thin layer drying model for RF-assisted heat pump drying of Ganoderma lucidum, in which, drying parameters included drying temperature of 40, 45 and 50°C and RF power of 0.65, 1.3 and 1.95 kW. The results indicated that the effective moisture diffusivity value increased with an increase in RF power that reduced the drying time by 10% and 21% at RF power of 1.95 kW as compared to RF power of 1.3 kW and 0.65 kW.

Nguyen Hay et al., (2022) [3] studied the RF-assisted heat pump drying of Ganoderma lucidum, in which, the drying parameters included drying air temperature of 40°C, air velocity of 1.2 m/s, and an RF power of 1.95 kW, 0.65 and 0 kW. The results showed that the increasing in RF power improved the heating rate and shortened the drying time significantly. In RF-assisted heat pump drying, the dried material got a uniform moisture distribution and temperature distribution and the color of dried products also maintained with the original red-brown color of fresh Ganoderma lucidum.

Based on the results of the previous studies mentioned above, RF heating technique is a suitable and effective choice for drying technology, in which, RF is combined with different drying methods to dry different types of materials. The advantage of RF volumetric heating mechanism can improve the heating rate, drying rate and quality of dried products both in terms of sensory quality and nutritional quality as well as the shelf life of dried products. In particular, the combination between RF with other drying methods as hot-air, vacuum and heat pump drying could reduce the drying time by 67% and the dried products got the higher quality as color, surface appearance and high ingredient retention of 98% as compared with the drying methods without RF assistance. However, the drying method using RF has to invest a high-cost equipment and the safety in operating RF drier must be considered as RF drier operates at high electric voltage.

4. Conclusions

With the increasing requirements of the agricultural and food processing industry, product quality in terms of sensory quality and nutritional quality as well as long storage time would play a crucial role. Besides, the cost of energy consumption also plays a significant role in reducing the processing cost and the price of dried products. The studies for finding out a suitable and the most effective drying technique has been, is and will be done. In particular, the hybrid drying technique with the combination of traditional drying technologies and IR, MW and RF has created a breakthrough both in terms of technology and economy. The effective support of the heating mechanism of IR, MW and RF greatly improves the drying efficiency such as high heating rate, short drying time, high drying product quality as well as saving considerable energy consumption of the drying system. The research, application and development of these hybrid drying technologies will certainly create a breakthrough for the agricultural and food processing industry in the future.

Acknowledgements

The authors would like to thank Nong lam University, Ho Chí Minh City, Vietnam and Van Lang University, Vietnam for funding this work.

Conflict of interest

The authors declare no conflict of interest.

Nomenclature

FD	Freeze drying
FIR	Far-infrared
IR	Infrared radiation
LHA	Low humidity air
MW	Microwave
MIR	Mid-infrared
NIR	near-infrared
RF	Radio frequency

Author details

Nguyen Hay[1], Le Quang Huy[2] and Pham Van Kien[3*]

1 Nong Lam University, Ho Chi Minh City, Vietnam

2 Cao Thang Technica`l College, Ho Chi Minh City, Vietnam

3 Faculty of Automotive Engineering, School of Engineering and Technology, Van Lang University, Ho Chi Minh City, Vietnam

*Address all correspondence to: kien.pv@vlu.edu.vn

References

[1] Wei X, Lau SK, Stratton J, Irmak S, Subbiah J. Radiofrequency pasteurization process for inactivation of salmonella spp. and enterococcus faecium NRRL B-2354 on ground black pepper. Food Microbiology. 2019;**82**:388-397. DOI: 10.1016/j.fm.2019.03.007

[2] Mayor L, Sereno AM. Modelling shrinkage during convective drying of food materials: A review. Journal of Food Engineering. 2004;**61**:373-386

[3] Hay N. Pham Van Kien and Le Anh Duc, mathematical model of radio frequency assisted heat pump drying of Ganoderma Lucidum (Ganoderma Boninense). International Journal on Advanced Science, Engineering and Information Technology. 2022;**12**(2): 726-731. DOI: 10.18517/ijaseit.12.2.9441

[4] Hnin KK, Zhang M, Mujumdar AS, Zhu Y. Emerging food drying technologies with energy-saving characteristics: A review. Drying Technology. 2019;**37**(12):1465-1480. DOI: 10.1080/07373937.2018.1510417

[5] Jumah R. Modelling and simulation of continuous and intermittent radio frequency- assisted fluidized bed drying of grains. Food and Bioproducts Processing. 2005;**83**(3):203-210. DOI: 10.1205/fbp.04291

[6] Jin W, Mujumdar AS, Zhang M, Shi W. Novel drying techniques for spices and herbs: A review. Food Engineering Reviews. 2018;**10**(1):34-45. DOI: 10.1007/s12393-017-9165-7. doi: 10.1080/07373937.2018.1458735

[7] Sakare P, Prasad N, Thombare N, Singh R, Sharma SC. Infrared drying of food materials: Recent advances. Food Engineering Reviews. 2020;**12**:381-398

[8] Jin W, Zhang M, Shi W. Evaluation of ultrasound pretreatment and drying methods on selected quality attributes of bitter melon (Momordica charantia L.). Drying Technology. 2019;**37**(3):387-396. DOI: 10.1080/07373937.2018.1458735

[9] Liu Y, Zhang Y, Wei X, Wu D, Dai J, Liu S, et al. Effect of radio frequency-assisted hot-air drying on drying kinetics and quality of Sichuan pepper (Zanthoxylum bungeanum maxim.). LWT-Food Science and Technology. 2021;**147**(3):111572. DOI: 10.1016/j.lwt.2021.111572

[10] Orikasa T, Koide S, Okamoto S, Imaizumi T, Muramatsu Y, Takeda J, et al. Impacts of hot air and vacuum drying on the quality attributes of kiwifruit slices. Journal of Food Engineering. 2014;**125**:51-58. DOI: 10.1016/j.jfoodeng.2013.10.027

[11] Liu Y, Sun Y, Yu H, Yin Y, Li X, Duan X. Hot air drying of purple-fleshed sweet potato with contact ultrasound assistance. Drying Technology. 2017;**35**(5):564-576. DOI: 10.1080/07373937.2016.1193867

[12] Ratti C, Mujumdar AS. Handbook of Industrial Drying. Boca Raton, Florida, Us: CRC Press; 2006. p. 1312. DOI: 10.1201/9781420017618

[13] Jain D, Pathare PB. Selection and evaluation of thin layer drying models for infrared radiative and convective drying of onion slices. Biosystems Engineering. 2004;**89**(3):289-296

[14] Sakai N, Hanzawa T. Applications and advances in farinfrared heating in Japan. Trends of Food Science Technology. 1994;**5**:357-362

[15] Fasina O, Tyler B, Pickard M, Zheng GH, Wang N. Effect of infrared heating on the properties of legume seeds. International Journal of Food Science & Technology. 2001;**36**:79-90

[16] Datta AK, Ni H. Infrared and hot-air-assisted microwave heating of foods for control of surface moisture. Journal of Food Engineering. 2002;**51**:355-364 http://onlinelibrary.wiley.com/journal/10.1111/(ISSN)1745-4530

[17] Bal S, Wratten FT, Chesness JL, Faulkner MD. An analytical and experimental study of radiant heating of rice grain. Transactions of the ASABE. 1970;**13**:644-647

[18] Erdogdu SB, Eliasson L, Erdogdu F, Isaksson S, Ahrne L. Experimental determination of penetration depths of various spice commodities (black pepper seeds, paprika powder and oregano leaves) under infrared radiation. Journal of Food Engineering. 2015;**161**:75-81

[19] Pawar SB, Pratape VM. Fundamentals of infrared heating and its application in drying of food materials: A review. Journal of Food Process Engineering. 2017;**40**:12308

[20] Sandu C. Infrared radiative drying in food engineering: A process analysis. Biotechnology Progress. 1986;**2**:109-119

[21] Mao Y, Wang S. Recent developments in radio frequency drying for food and agricultural products using a multi-stage strategy: A review. Critical Reviews in Food Science and Nutrition. 2021;**20**:1-18. DOI: 10.1080/10408398.2021.1978925

[22] Rastogi NK. Recent trends and developments in infrared heating in food processing. Critical Reviews in Food Science and Nutrition. 2012;**52**:737-760

[23] Venkitasamy C, Zhu C, Brandl MT, Niederholzer FJ, Zhang R, McHugh TH, et al. Feasibility of using sequential infrared and hot air for almond drying and inactivation of enterococcus faecium NRRL B-2354. LWT Food Science and Technology. 2018;**95**:123-128

[24] Yan JK, Wu LX, Qiao ZR, Cai WD, Ma H. Effect of different drying methods on the product quality and bioactive polysaccharides of bitter gourd (Momordica charantia L.) slices. Journal of Food Chemistry. 2019;**271**:588-596

[25] Zare D, Naderi H, Ranjbaran M. Energy and quality attributes of combined hot- air/infrared drying of paddy. Drying Technologyogy. 2015;**33**:570-582

[26] Onwude DI, Hashim N, Abdan K, Janius R, Chen G. The effectiveness of combined infrared and hot-air drying strategies for sweet potato. Journal of Food Engineering. 2019;**241**:75-87

[27] Qu F, Zhu X, Ai Z, Ai Y, Qiu F, Ni D. Effect of different drying methods on the sensory quality and chemical components of black tea. LWT- Food Science and Technology. 2019;**99**:112-118

[28] Wu XF, Zhang M, Bhandari B. A novel infrared freeze drying (IRFD) technology to lower the energy consumption and keep the quality of Cordyceps militaris. Innovative Food Science and Emerging Technologies. 2019;**54**:34-42

[29] Xie L, Mujumdar AS, Fang XM, Wang J, Dai JW, Du ZL, et al. Far-infrared radiation heating assisted pulsed vacuum drying (FIR-PVD) of wolfberry (Lycium barbarum L.): Effects on drying kinetics and quality attributes. Food and Bioproducts Processing. 2017;**102**:320-331

[30] Wanyo P, Siriamornpun S, Meeso N. Improvement of quality and antioxidant

properties of dried mulberry leaves with combined far-infrared radiation and air convection in Thai tea process. Food and Bioproducts Processing. 2011;**89**:22-30

[31] Chen Q, Bi J, Wu X, Yi J, Zhou L, Zhou Y. Drying kinetics and quality attributes of jujube (Zizyphus jujuba miller) slices dried by hot-air and short- and medium-wave infrared radiation. LWT- Food Science and Technology. 2015;**64**:759-766

[32] Orikasa T, Koide S, Okamoto S, Togashi C, Komoda T, Hatanaka S, et al. Temperature dependency of quality change during far-infrared drying of Komatsuna leaves. Acta Horticulturae. 2015;**1091**:319-325

[33] Giri SK, Prasad S. Drying kinetics and rehydration characteristics of microwave-vacuum and convective hot-air dried mushrooms. Journal of Food Engineering. 2007;**78**:512-521

[34] Salehi F, Kashaninejad M. Mass transfer and color changes kinetics of infrared-vacuum drying of grapefruit slices. International Journal of Fruit Science. 2018;**18**:394-409

[35] Salehi F, Kashaninejad M. Modeling of moisture loss kinetics and color changes in the surface of lemon slice during the combined infrared-vacuum drying. Information Processing in Agriculture. 2018;**5**:516-523

[36] Khampakool A, Soisungwan S, Park SH. Potential application of infrared assisted freeze drying (IRAFD) for banana snacks: Drying kinetics, energy consumption, and texture. LWT. 2019;**99**:355-363

[37] Antal T. Comparative study of three drying methods: Freeze, hot air-assisted freeze and infrared-assisted freeze modes. Agronomy Research. 2015;**13**:863-878

[38] Pääkkönen K. A combined infrared/heat pump drying technology applied to a rotary dryer. Agricultural and Food Science. 2002;**11**(3):209-218. DOI: 10.23986/afsci.5726

[39] Khodifad BC, Dhamsaniya NK. Drying of food materials by microwave energy - A review. International Journal of Current Microbiology and Applied Sciences. 2020;**9**(5):1950-1973

[40] Soysal Y. Microwave drying characteristics of parsley. Journal of Biosystems Engineering. 2004;**89**(2):167-173

[41] Yanyang X, Min Z, Mujumdar AS, Le-qun Z, Jin-cai S. Studies on hot air and microwave vacuum drying of wild cabbage. Drying Technologyogy. 2004;**22**(9):2201-2209

[42] Therdthai N, Zhou W. Characterization of microwave vacuum drying and hot air drying of mint leaves (Mentha cordifolia Opiz ex Fresen). Journal of Food Engineering. 2009;**91**(3):482-489

[43] Wang R, Zhang M, Mujumdar AS, Sun JC. Microwave freeze-drying characteristics and sensory quality of instant vegetable soup. Drying Technology. 2009;**27**(9):962-968

[44] Patil GD, Pardeshi IL, Shinde KJ. Drying of green leafy vegetables using microwave oven dryer. Journal Ready to Eat Food. 2015;**2**:18-26

[45] Akal D, Kahveci K. Investigation of microwave drying characteristics of carrot slices. In: Proceedings of the 2nd World Congress on Mechanical, Chemical, and Material Engineering (MCM'16); August 22-23. Budapest,

Hungary: HTFF; 2016. pp. 112-1-112-5. DOI: 10.11159/htff16.112

[46] Horuz E, Bozkurt H, Karataş H, Maskan M. Effects of hybrid (microwaveconvectional) and convectional drying on drying kinetics, total phenolics, antioxidant capacity, vitamin C, colour and rehydration capacity of sour cherries. Food Chemistry. 2017;**230**:295-305

[47] Deepika S, Sutar PP. Combining osmotic-steam blanching with infrared–microwave–hot air drying: Production of dried lemon (Citrus Limon L.) slices and enzyme inactivation. Drying Technology. 2018;**36**(14):1719-1737

[48] Rodriguez A, Bruno E, Paola C, Campañone L, Mascheroni RH. Experimental study of dehydration processes of raspberries (RubusIdaeus) with microwave and solar drying. Food Science and Technology. 2019;**39**(2):336-343

[49] Jinwoo Park, Jae Hyun Cho, Richard D. Braatz, Mathematical modeling and analysis of microwave-assisted freeze-drying in biopharmaceutical applications. Computers and Chemical Engineering 2021;**153**:107412

[50] Munoz I, Serra X, Guardia MD, Fartdinov D, Arnau J, Picouet P, et al. Radio frequency cooking of pork hams followed with conventional steam cooking. LWT-Food Science and Technology. 2020;**123**:109104. DOI: 10.1016/j.lwt.2020.109104

[51] Palazoğlu TK, Miran W. Experimental comparison of microwave and radio frequency tempering of frozen block of shrimp. Innovative Food Science & Emerging Technologies. 2017;**41**:292-300. DOI: 10.1016/j.ifset.2017.04.005

[52] Liao M, Damayanti W, Xu Y, Zhao Y, Xu X, Zheng Y, et al. Hot air-assisted radio frequency heating for stabilization of rice bran: Enzyme activity, phenolic content, antioxidant activity and microstructure. LWT-Food Science and Technology. 2020;**131**:109754. DOI: 10.1016/j.lwt.2020.109754

[53] Pegna FG, Sacchetti P, Canuti V, Trapani S, Bergesio C, Belcari A, et al. Radio frequency irradiation treatment of dates in a single layer to control Carpophilus hemip terus. Biosystems Engineering. 2017;**155**:1-11. DOI: 10.1016/j.biosystemseng.2016.11.011

[54] Lin Y, Subbiah J, Chen L, Verma T, Liu Y. Validation of radio frequency assisted traditional thermal processing for pasteurization of powdered infant formula milk. Food Control. 2020;**109**:106897. DOI: 10.1016/j.foodcont.2019.106897

[55] Liao M, Zhao Y, Gong C, Zhang H, Jiao S. Effects of hot air-assisted radio frequency roasting on quality and antioxidant activity of cashew nut kernels. LWT-Food Science and Technology. 2018;**93**:274-280. DOI: 10.1016/j.lwt.2018.03.047

[56] Chen L, Subbiah J, Jones D, Zhao Y, Jung J. Development of effective drying strategy with a combination of radio frequency (RF) and convective hot-air drying for inshell hazelnuts and enhancement of nut quality. Innovative Food Science & Emerging Technologies. 2021;**67**:102555. DOI: 10.1016/j.ifset.2020.102555

[57] Roknul ASM, Zhang M, Mujumdar AS, Wang Y. A comparative study of four drying methods on drying time and quality characteristics of stem lettuce slices (Lactuca sativa L.). Drying Technology. 2014;**32**(6):657-666. DOI: 10.1080/07373937.2013.850435

[58] Zhou X, Xu R, Zhang B, Pei S, Liu Q, Ramaswamy HS, et al. Radio frequency-vacuum drying of kiwifruits: Kinetics, uniformity, and product quality. Food and Bioprocess Technology. 2018;**11**(11):2094-2109. DOI: 10.1007/s11947-018-2169-3

[59] Zhou X, Gao H, Mitcham EJ, Wang S. Comparative analyses of three dehydration methods on drying characteristics and oil quality of in-shell walnuts. Drying Technology. 2018;**36**(4):477-490. DOI: 0.1080/07373937.2017.1351452

[60] Zhou X, Ramaswamy H, Qu Y, Xu R, Wang S. Combined radio frequency-vacuum and hot air drying of kiwifruits: Effect on drying uniformity, energy efficiency and product quality. Innovative Food Science & Emerging Technologies. 2019;**56**:102182. DOI: 10.1016/j.ifset.2019.102182

[61] Zhang H, Gong C, Wang X, Liao M, Yue J, Jiao S. Application of hot air-assisted radio frequency as second stage drying method for mango slices. Journal of Food Process Engineering. 2019;**42**(2):e12974. DOI: 10.1111/jfpe.12974

[62] Ran X, Zhang M, Wang Y, Liu Y. Vacuum radio frequency drying: A novel method to improve the main qualities of chicken powders. Journal of Food Science and Technology. 2019;**56**(10):4482-4491. DOI: 10.1007/s13197-019-03933-0

[63] Peng J, Yin X, Jiao S, Wei K, Tu K, Pan L. Air jet impingement and hot air-assisted radio frequency hybrid drying of apple slices. LWT-Food Science and Technology. 2019;**116**:108517. DOI: 10.1016/j.lwt.2019.108517

[64] Wang W, Wang Y, Yang R, Tang J, Zhao Y. Hot-air assisted continuous radio frequency heating for improving drying efficiency and retaining quality of inshell hazelnuts (Corylus avellana L. cv. Barcelona). Journal of Food Engineering. 2020;**279**:109956. DOI: 10.1016/j.jfoodeng.2020.109956

[65] Wang C, Kou X, Zhou X, Li R, Wang S. Effects of layer arrangement on heating uniformity and product quality after hot air assisted radio frequency drying of carrot. Innovative Food Science & Emerging Technologies. 2021;**69**:102667. DOI: 10.1016/j.ifset.2021.102667

[66] Shewale SR, Rajoriya D, Bhavya ML, Hebbar HU. Application of radiofrequency heating and low humidity air for sequential drying of apple slices: Process intensification and quality improvement. LWT-Food Science and Technology. 2021;**135**:109904. DOI: 10.1016/j.lwt.2020.109904

[67] Duc LA, Hay N, Van Kien P. Mathematical model of thin layer drying of ganoderma lucidum by radio frequency assisted heat pump drying. Frontiers in Heat and Mass Transfer. 2022;**18**(44):1-7. DOI: 10.5098/hmt.18.44

Chapter 4

Role of Food Microwave Drying in Hybrid Drying Technology

Bandita Bagchi Banerjee and Sandeep Janghu

Abstract

Dehydration is the key to food preservation reducing volume and increasing shelf life. Dehydration technology has witnessed renaissance with the development of advanced technology such as microwave drying, freeze drying, fluidized bed drying, and refractance window drying. Combination of drying methods has increased the versatility of dehydration process of which field-based drying methods have always been hyped and microwave drying being the most adorned of all, considering its ease of fabrication and drying efficiency. Synergizing it with methods such as hot air drying, freeze drying, fluidized bed drying, or vacuum drying enhances its performance and the quality of the dried product. The merits and functionality of each method in hybrid drying with microwave have been discussed in the chapter.

Keywords: drying, moisture, microwave, hybrid, efficiency

1. Introduction

Dehydration is proclaimed as the key to increased shelf life of perishable food, pertaining to the reduction of available moisture for microbial growth. It has been one of the most ancient apt techniques for food preservation, the conventional methods being air drying, solar drying, etc. Dehydration has not only contributed to the extended life of fresh food and throughout the year availability of seasonal products [1] but also has minimized the hassle of voluminous handling of raw and processed products, thus paving its way to reach end consumers like space researchers in spaceships, marine staff, war zones, and needy ones in far flung inaccessible areas.

Researchers and scientists have experimented with dehydration methods such as fluidized bed drying, freeze drying, osmotic dehydration, drying by natural radiation, spray drying, to understand the compatibility of the methods with the different kinds of food in terms of nutrient retention, operation cost, efficiency, shelf life, and final quality of the dehydrated product. Each method or device projected for dehydration has evolved with time. The relentless efforts to attain maximum efficiency have paved way for the advanced dehydration methods and gradually have ignited the concept of combining the individual methods, hence introducing the hybrid drying technology. The amalgamation of two to more drying techniques has also added to the versatility of the methods. The demand to adhere to the norms and policies in relation to environment, government regulation, increased throughput, quality, etc., has necessitated enumeration of methodologies such as hybrid drying technique for achieving optimum results. The conventional convective hot air drying has been improvised many times by researchers with combination of nonthermal drying techniques such as ultrasound, ultraviolet radiation, or pulse electric field [2], which proved to produce better quality dried products in lesser drying time compared to single-operation system of hot air drying. Subsequently, the prospects of combining the nonthermal technologies with other kinds of drying techniques have been explored by thinkers. Of all methods, field-based drying methods have received ample attention by researchers due to the profound advantages of these techniques such as improved drying kinetics, efficient rehydration ratio, and high-quality end product. The promising scope of synergized drying methods using field-based methods has encouraged experimentation with different combination sets of which microwave drying has been popular, due to the ease of fabrication, nutrient retention of food, and cost effectivity. Hybrid dehydration processes with microwave treatment envisage to minimize loss of heat-sensitive food components. This chapter focusses on microwave drying hybridization with other drying methods, thus compounding an insight herein on the breaches attended and the lags yet to be addressed through further research.

2. Advancement in dehydration technology

The ardent efforts on improvisation of dehydration technique and application to different kinds of foods have been relentless, thus nearing perfection from all aspects inch by inch. The traditional sun-drying method inspired designing of solar dryers by intervention of science and technology. Conventional air drying method accentuated with the introduction of hot air dryers and oven dryers to reduce dependency on natural sources. However, the process being energy-driven involved huge amount of recurring cost due to the immense consumption of high-priced electric energy, thus subsequently bringing in novel technologies such as vacuum drying, freeze drying, spray drying, and ample of advanced methods, which enhanced the drying behavior further with comparatively lesser energy need [1]. Energy efficiency of the drying method has always been the focus point to optimize the operation cost without compromising on product quality. The heat pump dehumidifiers and superheated steam drying proved to be energy-efficient methods, resulting in dried products with better rehydration properties and nutritional content. The heat pump dehumidifiers enabled recycling of air by adding and rejecting heat during the process through heat pumps and condensers resulting in minimization of energy consumption. Superheated steam drying has shown promising results in terms of retention of heat-sensitive compounds of food such as vitamin C, thus encouraging its implementation in the food industry [3–5]. Microwave drying is known for its deep penetration characteristic within food layers and plays an important role in food drying, especially when in combination with other drying methods. Microwave vacuum drying is reported to be efficient at 26–52°C. When subjected at 640–710 W to orange juice concentrates for the production of fruit gel, the color of the gel has been attractive and lighter as compared to the ones prepared by air drying [6]. Plate-transducer power ultrasonic generators aided the generation of high-intensity ultrasound waves, which added new dimension to dehydration of vegetables. Several techniques on application of high-intensity ultrasound waves directly to vegetable surfaces have been explored by scientists as ultrasounds increase the evaporation rate of moisture substantially reducing the duration of drying [7]. Refractance window drying system has gained popularity over the years because of its efficiency in terms of quality retention and inexpensive equipment setup. It has shown impressive performance in the conversion of liquid products such as juice, puree, to flakes, leather, sheets, etc., having attractive color, aroma, and antioxidant activity. It is used for the production of egg powder, fruit powder, herbal extract, etc. [8]. Nonthermal drying technologies like osmotic drying have been effective for fruits like apricot with an initial pretreatment of drying material. Dielectric drying through radio frequency (RF) and microwave systems has been extensively used in food industry, wherein electrical energy is converted to heat energy by polarizing the electric dipoles in the food material with electric fields. The techniques are apt enough to be successfully replicated for other kinds of high-value fruits as studied in apricot drying [9].

3. Advancement in hybrid dehydration methods

Hybrid methods of drying are prevalent to envisage the merits at the maximum extent and have a versatile approach toward efficient drying. Combination of two or three drying methods not only reduces the sole dependency on hot air drying but also increases energy efficiency and better nutritional retention in comparison with the hot air-dried products that are more susceptible to nutritional loss, shrinkage, color change, flavor change, and texture hardness [10]. The approach of hybrid drying thrives to conglomerate the merits of individual drying techniques, minimizing the demerits of each. Coupling of convective drying with field-based drying methods has been reported to be effective to retain nutritional and functional components of fruits and vegetables. Application of ultrasounds is a nonthermal method of dehydration. Subjecting apple slices to ultrasonication after convective air drying has shown reduced processing time and better quality retention [11]. Infrared radiation is also implemented to heat-sensitive food products for drying in a synergistic way with freeze drying and vacuum drying. It is being used for drying of high-value herbs such as *Ginkgo biloba* and *Cordyceps militaris* [12]. Microwave drying has been the most sought after dehydration method to synergize with other methods. Electromagnetic waves penetrate food, causing oscillation of molecules which in turn generate heat inside the food. The technology coupled with other drying methods increase drying efficiency manifold. For drying of spices, microwave drying has proved to be effective when conjugated with drying methods such as infrared drying or fluidized bed drying [13].

4. Hybrid drying with field-based drying methods

The field-based drying methods are the ones that employ electromagnetic energy and acoustic energy such as microwave, infrared radiation, radio frequency (RF), and ultrasound for nonthermal drying. These kinds of drying techniques exhibit better food quality in terms of appearance and nutrition value. In addition, the techniques have rapid drying kinetics and improved thermal efficiency in comparison with the conventional hot air drying. Individually, each technique has its own set of advantages. Introduction of field-based drying methods to the conventional techniques imbibes its own benefits such as increased drying efficiency and low energy consumption.

4.1 Benefits of field-based drying technology

4.1.1 Microwave drying

i. Generates internal vapor

ii. Heating is volumetric

iii. Drying rate is high

4.1.2 Drying with radio frequency

i. Wavelength employed is longer than microwave

ii. Penetrates the food deep

4.1.3 Infrared drying

i. Transfer medium is not required for energy deliver

ii. Specific areas on food surface can be targeted

4.1.4 Ultrasonic drying

i. Strongly adhered moisture can be removed [10]

5. Hybrid drying with microwaves

Fabrication of field-based drying models calls for considerable technical inputs and high costs. Sophisticated equipment setups are required to run the drying system [14]. Maintenance and running of the hybrid setups with field-based drying requires technical knowledge and expertise. Considering the ease of modeling and feasibility of hybrid drying with field-based methods, microwave drying is best suited for miniature setups as well as extensive layouts. However, microwave drying if employed solely lacks energy efficiency, because removal of water using microwaves is costly considering the high rates of electric energy. Moreover, useful heat is not generated from all the emitted microwaves [15]. Microwave drying shows best result when combined with other formats of drying.

5.1 Heat features of microwave

The operating frequency of microwave is 300 MHz to 300 GHz. The domestic microwave oven induces microwaves of frequency 2450 MHz, whereas in commercial large-scale level microwaves of range 900 MHz are deployed.

The heating mechanism involves absorption of microwave energy by the water present in the food cavities, resulting in evaporation of water due to temperature rise. Microwave heating is a dielectric heating system, wherein the molecules carrying negative and positive charge (such as sugar, fat, and water molecules) align themselves to the alternating electric field subjected by the microwaves. The process in turn generates heat as the rotating molecules strike other molecules and set them in motion causing difference in vapor pressure between the central and surface layers of food. This causes moisture in the food to travel out of it fast making microwave drying a rapid, uniform, and energy complacent method in comparison with hot air drying. Penetration of microwaves into food depends on the composition of food and also the frequency of microwave. The microwaves with lower frequency penetrate deeper into the food in comparison with the ones of higher frequencies.

The electromagnetic energy is converted to heat energy for temperature rise. Hence, the heat generation rate per unit volume Q determines the rate of temperature increase. The Q value can be enumerated as follows [16]:

$$Q = 2\pi f \varepsilon_0 \varepsilon'' E^2_{rms} \tag{1}$$

where Q = heat generated per unit volume.

f = frequency

ε_0 = 8.854 x 10–12 F/m (free space permittivity)

ε'' = material's dielectric loss factor

E_{rms} = root mean square of electric field intensity at the location

The temperature increases as time increases for any given location in the food which is expressed as follows [16]:

$$Q = \rho C_p \frac{\Delta T}{\Delta t} \tag{2}$$

where ρ = material density

Cp = specific heat capacity

T = temperature

t = time

Replacing Eq. (1) with Eqs. (2) and (3) is obtained [16]:

$$\rho C_p \frac{\Delta T}{\Delta t} = 2\pi f \varepsilon_0 \varepsilon'' E^2_{rms} \tag{3}$$

From Eq. 3, the change in temperature with respect to time can be calculated for any location of food subjected to microwave heating [16, 17].

5.2 Microwave with hot air drying

Hot air drying in itself is a time-consuming and energy-demanding process. The extensive exposure to high temperature causes loss of nutrients and heat-sensitive compounds of the food. Migration of solutes to food surface occurs resulting into case hardening. Rehydration of the dehydrated food becomes a challenge due to shrinkage of the food during drying. Microwaves in conjunction with hot air drying minimize the drawbacks of both kinds of method. Application of this hybrid format of drying is gaining popularity in the industries and factories. The drying time is considerably reduced with accelerated energy flow to the evaporating point of the food structure even when in moving state within the product. The microwave drying drives moisture from the center to the surface from where it is deliberately removed by the airflow from hot air dryer. The moisture carrying capacity of the airflow is determined by its velocity and temperature that need to be controlled keeping in consideration the heat sensitivity of the volatile components and phytochemicals present in the product. Herein, optimization of airflow, temperature, and microwave power density is crucial for success of the process [15].

Apart from continuous microwave drying, intermittent microwave drying is also gaining popularity for its efficient performance with convective hot air drying. It overcomes the drawbacks of continuous microwave drying such as humidity and uneven temperature distribution, through the power-off phases in the working cycles during which there is a chance of even distribution of moisture and heat by the continuous hot airflow. This subsequently enhances the drying characteristic of intermittent microwave heating in comparison with continuous microwave in terms of nutrient retention and quality standards, though the time involved is longer in intermittent microwave heating when coupled with hot air drying [14]. Food quality might be susceptible to constant microwave heating for long duration which is, however, not so as in case of subjection to intermittent microwaves. Studies on osmotic dehydrated strawberries showed that intermittent microwave drying with magnetron at switched off condition for almost 74% of the time enumerated high-quality strawberries in terms of retention of phenolics, antioxidants, and anthocyanins [18]. Similarly in case of parsley subjected to hybrid drying conditions such as convective hot air microwave drying and intermittent microwave drying showed that retention of vitamin C, color, and rehydration property was the best for the ones with intermittent microwave drying [19].

5.3 Microwave with freeze drying

Freeze drying is the freezing of moisture in food to ice and subsequent conversion of ice to vapor on lowering the pressure. Freeze drying by itself is an energy consuming method, hence subjected to high-value products such as cocoa, coffee beans, and nutraceutical herbs. The dehydration rate is low, and cost incurred is high. In spite of being cost-intensive, it is the most sought after drying method for heat-sensitive products because of its acclaimed capacity to conserve the macronutrients, micronutrients, color, antioxidative properties, and functional properties of the food. The advantage of the method is the sublimation stage of the ice crystals without involvement of oxygen, which enables successful retention of the quality attributes of a product such as structure, color, flavor, and aroma and help easy rehydration [10].

Combining freeze drying with microwave drying process accentuates its efficacy by minimizing its disadvantages in terms of time duration, energy efficiency, and cost. Conventional freeze drying occurs by transfer of energy subsequently from the dried layer to the low-conductive frozen layer through the bulk of the food involving long drying duration and energy requirement [20]. Microwaves when employed together with freeze drying reduce the drying time as compared to freeze drying [21] due to the volumetric heating by microwaves. As ice has low loss factor, microwave energy penetrates only the organic portion of the frozen food, thus warming up all regions of the food instead of layer-by-layer heating in case of conventional freeze dry method. The merits of microwave freeze drying are enumerated as follows:

a. Microwaves penetrate deep into the frozen product

b. The energy is dissipated rapidly through the food material

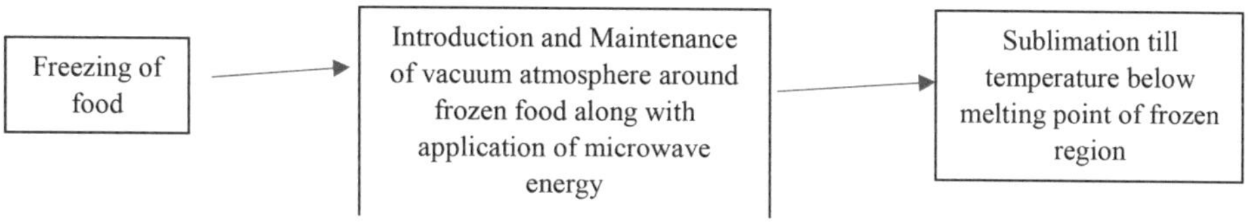

Figure 1.
Steps in hybrid dying by microwave freeze drying [20].

c. Microwaves adapting automatically to the products dynamic dielectric property

d. Less of energy required during efficient drying at the falling rate stage [22]

Microwaves in freeze drying can be subjected in two ways:

a. Simultaneously microwave drying and freeze drying of food wherein the microwave is applied right from the onset of the freeze drying occurring in vacuum condition, to provide necessary energy for sublimation.

b. Two-phased drying initiated with freeze drying followed by microwave drying. Herein, freezing of the product is first completed followed by introduction of microwave energy at vacuum condition in the storage chamber containing the dried product. As the energy from microwaves heat the frozen food, its bulk temperature starts increasing and the frozen water gets evaporated directly to gaseous state escaping in the vacuum chamber. The transition creates an interface between frozen part and dried part, which gradually reduces with time to finally obtain a uniformly dried product (**Figure 1**)

Wang and Chen (2007) [21] proposed use of dielectric bars or spheres for microwave freeze drying of materials lower in solid content or solid products having low loss factor, thus demonstrating 20% reduction in drying time. Skim milk was microwave freeze-dried using silicon carbide as the dielectric material, wherein the drying time was significantly lesser then freeze drying. Duan et al. 2007 [22] reported application of microwave freeze drying to cabbages, which exhibited lower drying time and sterilization effect. Similar was the finding by Duan and Zhang (2008) [23], wherein drying time was reduced to almost half in comparison with conventional freeze drying of sea cucumbers.

5.4 Microwave fluidized bed drying

The prime drawback faced during microwave drying is the nonuniform drying of the product at a given time. Such kind of demerit is overcome by combining fluidized bed drying to the microwave process. In fluidized drying process, the food to be dried is subjected to hot air at high pressure through a porous bed causing agitation to the food particles at which if microwave is introduced, each particle will receive the radiation leading to uniform heating and a subsequent reduction in the diffusion time of drying [24]. The disadvantage of fluidized bed drying is the undesirable size reduction of food particles due to the collision among the particles. Hence, drawbacks of both kinds of drying techniques are minimized in the hybrid drying as described by

Goksu et al. (2004) [25] who studied the drying pattern of macaroni beads subjected to fluidized bed and microwave drying. On comparison of drying time, it was observed that it reduced by 50% when both the methods were applied synergistically with 2.1 and 3.5 W/g of microwave energy in a fluidized bed dryer as against application of microwave and fluidized bed drying individually. Rehydration capacity is another determining factor to ensure the success of a given drying process. Khoshtaghaza et al. (2015) [24] observed that microwave-fluidized drying of soyabeans had better rehydration ratio when subjected to high power of microwave and low velocity of air, because expansion of cell walls occurred as its elasticity and starch hydration increased due to the applied heat and high pressure generated internally in the kernels by the microwave energy.

The selection of microwave energy and temperature of hot air in fluidized bed drying is crucial to obtain the ideal combination for energy efficiency and uniform drying. On application of high-power microwave, the temperature of the particles on surface will be higher than the air temperature of fluidized bed dryer; thus, airflow cools the particles instead of evaporating moisture. Taheri et al. (2019) [26] designed a fluidized bed microwave dryer, as depicted in **Figure 2**, to study the drying curves and moisture diffusivity of lentil seed at different exposure time. Drying was experimented with 0, 300, 400, and 500 W microwave powers at 50 and 60°C fluidized hot air, which aided disinfection along with drying of lentil seeds. Moisture diffusivity varied from 0.44×10^{-10} (when subjected only to air heated to 50°C) to 3.06×10^{-10} m^2/s (when microwave power of 500 W is used with air heated to 60°C) by considering convective boundary condition for the seeds as per Fick's second law of diffusion [26].

5.5 Microwave with vacuum drying

The volume heating in case of vacuum drying is apt for bulk-drying of products with low thermal conductivity and high heat sensitivity such as the viscous and

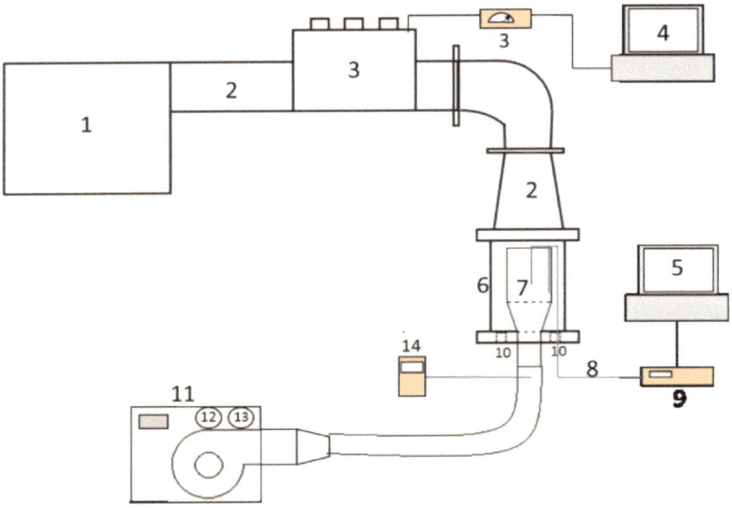

Figure 2.
Schematic diagram of fluidized bed microwave dryer [26]. Where 1 = magnetron, 2 = waveguide, 3 = tuner, 4 = power monitor, 5 = data and microwave power control system, 6 = microwave cavity, 7 = sample holder, 8 = fiber optic probes, 9 = temperature monitor, 10 = vents, 11 = airblower, 12 = air speed potentiometer, 13 = air temperature potentiometer, 14 = inlet air temperature monitor.

sugar-rich products [27]. In vacuum, moisture evaporates at temperatures lower than that at atmospheric pressure, thus protecting the products from high temperatures. As air is not involved in this format of drying, incidence of oxidation reactions is eliminated which helps to keep the taste, flavor, color, etc., intact [28] favoring its application in horticultural produces. The advantages of vacuum drying if imbibed into the microwave drying method are the merits of the hybrid model is manifold in terms of rapid heating, energy efficiency, quality retention, and uniform heating. The moisture in food is rapidly heated to vaporize at the requisite low temperature of vacuum environment by developing microwave-induced electric field on the food's water molecules [29].

The benefits of the process are almost equal to freeze drying method in terms of minimization of nutrient loss, structural and flavor change due to short heating duration at low temperatures, but with comparatively lower cost indulgence. Moreover, structural collapse is reduced with the development of porous structure as the moisture migrates from the high-pressure core zone of the food, developed by microwave heating, to the vacuum zone surrounding the material, the phenomenon being referred to as "puffing" [30]. Efficiency of a microwave vacuum dryer calls for consideration on the following details:

a. The dielectric property of sample

b. Distribution of microwave energy homogeneously through the cross section of sample bed

c. The rate of product throughput

d. Vacuum depth

e. Use of DC microwaves to avoid peaks in electric fields as vacuum reduces breakdown field strength, hence, if exceeded may result plasma or sparks

The microwave vacuum dryers can be either fabricated as static dryers or rotary dryers. The static dryers are similar to the domestic microwave oven with an additional vacuum chamber. The design is cost-effective and efficient for thin-layer products. The basic schematic representation of the design is represented in **Figure 3** [31, 32]. With samples having multiple layers, the design is challenging as drying duration increases due to the long time taken by microwaves to penetrate the layers causing

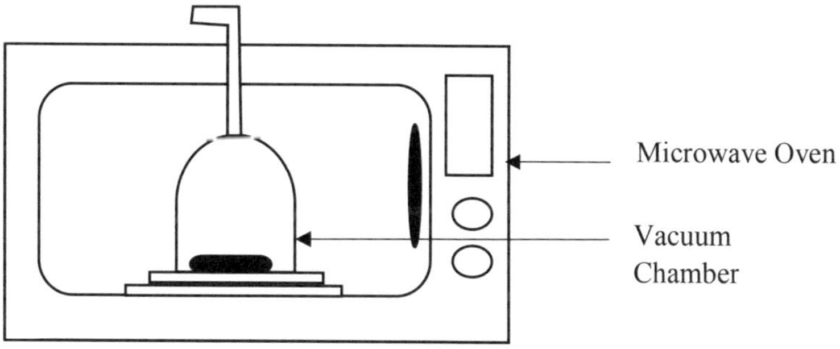

Figure 3.
Schematic diagram of static microwave vacuum dryer [31, 32].

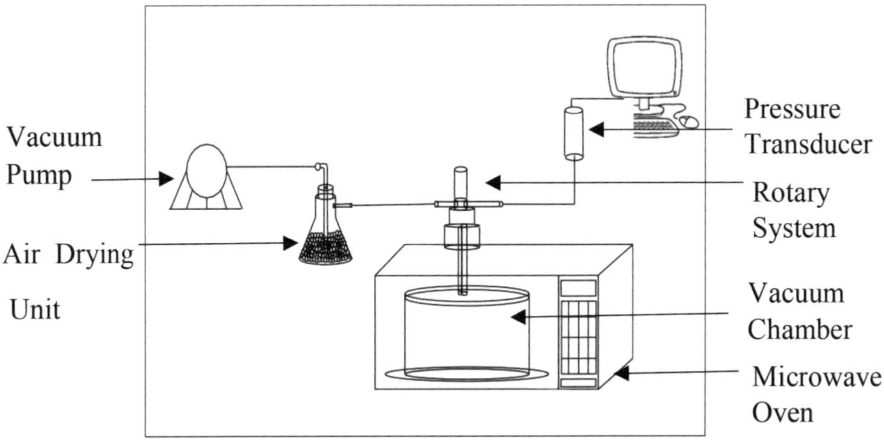

Figure 4.
Schematic diagram of rotary microwave vacuum dryer [27].

nonuniform temperature spectrum. The disadvantage of static dryers have given way to the development of rotary microwave vacuum dryers, wherein a rotating mechanism is introduced for rotation of the cylindrical basket or high HDPE drum inside the microwave oven. The rotational speed of the rotary system is controlled to avoid electric arc damage and ensure uniform heating of drying sample through homogeneous dissemination of microwave energy, which penetrates the multiple layers of food diligently [33]. The schematic representation of the design is represented in **Figure 4** [27].

This hybrid method of drying has shown impressive results. Monteiro et al. (2015) [27] designed the model as in **Figure 4** and experimented with samples such as grapes, banana, tomato, and carrot slices to obtain dried products close to the quality of freeze-dried products within 20 minutes as against 14–16 hours in freeze drying. Cranberries dried in microwave vacuum system retained the antioxidant activity and the bioactive compounds, which is not the case when dried solely with microwaves [34]. Drying process of garlic granules from 21 percent moisture content to 3 percent was the most efficient by microwave vacuum drying in comparison with microwave hot air drying as related to drying rate and product temperature [35]. By implementing automatic temperature control in microwave vacuum drying of strawberry, moisture could be reduced to 6.85 and the rehydration achieved about 55% keeping color, texture, and flavor intact [28].

5.6 Advantages of microwave energy

The microwave heating has its own set of advantages, tapping of which is feasible by imbibing into other drying methods. The advantages of this volume heating can be glanced to note as follows.

1. A high partial pressure is developed in the product due to higher internal temperatures as compared to the surface, which drives the moisture out from within the material, thus enabling to maintain permeable surface layers instead of over drying of surfaces.

2. Deep penetration by microwaves

3. Selective heating of water and organic solvents because of waters' higher dielectric losses compared with other molecules

4. Efficient drying of high-moisture products with low thermal conductivity

5. Control of energy transport speed feasible

6. Automation feasible

7. Shorter processing duration

6. Conclusion

Microwave energy synergized with different drying methods augments the drying efficiency by tapping its adventitious characteristics. The positive features of the technique also neutralize its certain negative features such as nonuniform drying and high electric charge rates. Imbibing microwave drying into hybrid drying technology is not only a way to enhance the drying rates but also a novel approach to retain the quality of the product. Hot air drying accelerates the escape of moisture from the core to the surface and then to the environment, while freeze drying aids to conserve the bioactive compounds and nutritional status of the products. Fluidized bed drying and vacuum drying help attaining better rehydration ratio and uniform heating of the products. Hence, holistic approach is the key to a profound hybrid drying setup bringing in efficiency and quality together.

Hybrid drying has embarked and popularized in recent times due to its immense versatility and drying efficiency. Its potency will enhance further in the future time if there is more of studies on the combination of different drying methods in a single or multiple setup and if more of initiative is exhibited on the transfer of technology to implementation level which is the need of the hour. Minimization of the fabrication cost will boost the use of hybrid drying in small-scale food industries as well. The individual methods in hybrid drying should be selected considering the need to obtain quality finished product with minimum environment stress. The efficiency of the techniques can further enhance with the use of sensors to keep a real-time check on the quality, hence adding a new perspective to it. The deterioration of quality and degradation of heat-sensitive nutrients have always been cause of concern and deciding factors for judging the efficiency of drying methods. Hence, the implication of sensors is manifold for detecting deterioration of molecules such as ascorbic acid, phytochemicals, or other such heat-sensitive antioxidant compounds during runtime, which can subsequently be communicated to the operator or automatically process optimized through the use of appropriate software [36].

Acknowledgements

The authors are grateful to NIFTEM-T and its Liaison Office, Guwahati, for the support.

Conflict of interest

The authors declare no conflict of interest.

Author details

Bandita Bagchi Banerjee and Sandeep Janghu*
The National Institute of Food Technology, Entrepreneurship
and Management - Thanjavur (NIFTEM-T), Guwahati-LO, Assam, India

*Address all correspondence to: sandeep@iifpt.edu.in

References

[1] Jayaraman KS, Das Gupta DK. Dehydration of fruits and vegetables-recent developments in principles and techniques. Drying Technology. 1992; **10**(1):1-50

[2] Onwude DI, Hashim N, Janius R, Abdan K, Chen G, Oladejo AO. Non-thermal hybrid drying of fruits and vegetables: A review of current technologies. Innovative Food Science & Emerging Technologies. 2017;**43**: 223-238

[3] Kohayakawa MN, Silveria-Junior V, Telis-Romero J. Drying of mango slices using heat pump dryer. In: Proceedings of the 14th International Drying Symposium. Sao Paulo, Brazil, 22-25 August 2004

[4] Tatemoto Y, Yano S, Mawatari Y, Noda K, Komatsu N. Drying characteristics of porous material immersed in a bed of glass beads fluidized by superheated steam under reduced pressure. Chemical Engineering Science. 2007;**62**(1–2):471-480

[5] Mallik A, Arman AM, Kundu S. Drying and dehydration technologies: A compact review on advance food science. MOJ Food Process Technology. 2018;**6**(1):36-40. DOI: 10.15406/mojfpt.2018.06.00142

[6] Drouzas AE, Tsami E, Saravacos GD. Microwave/vacuum drying of model fruit gels. Journal of Food Engineering. 1999;**39**(2):117-122

[7] Gallego-Juárez JA, Riera E, Blanco SF, Rodríguez-Corral G, Acosta-Aparicio VM, Blanco A. Application of high-power ultrasound for dehydration of vegetables: Processes and devices. Drying Technology: An International Journal. 2007;**25**(11):1893-1901. DOI: 10.1080/07373930701677371

[8] Nindo CI, Tang J. Refractance window dehydration technology: A novel contact drying method. Drying Technology. 2007;**25**(1):37-48

[9] Saufishan TA, Harimuthiah S, Arokiyaraj AA, Arshiya C, Aravind G, Gandhi SS, et al. Recent advances in apricot dehydration. International Journal of Scientific & Engineering Research. 2020;**11**(10):640-647

[10] Hii CL, Ong SP, Yap JY, Putranto A, Mangindaan D. Hybrid drying of food and bioproducts: A review. Drying Technology. 2021;**39**(11):1554-1576

[11] Sabarez HT, Gallego-Juarez JA, Riera E. Ultrasonic-assisted convective drying of apple slices. Drying Technology: An International Journal. 2012;**30**(9):989-997

[12] Boateng ID, Yang XM, Li YY. Optimization of infrared-drying parameters for Ginkgo biloba L. seed and evaluation of product quality and bioactivity. Industrial Crops and Products. 2021;**160**:113108

[13] Jin W, Mujumdar AS, Zhang M, Shi W. Novel drying techniques for spices and herbs: A review. Food Engineering Reviews. 2018;**10**(1): 34-45

[14] Li K, Zhang M, Mujumdar AS, Chitrakar B. Recent developments in physical field-based drying techniques for fruits and vegetables. Drying Technology. 2019;**37**(15):1954-1973

[15] Dehnad D, Jafari SM. Combined microwave drying with other drying methods. In: The First Middle East Drying Conference: Shahid Chamran. Iran: University of Ahvaz; 2012

[16] Mirzabeigi Kesbi O, Rajabipour A, Omid M, Goldansaz SH. Determination of electric field intensity during microwave heating of selected vegetables and fruits. Journal of Microwave Power and Electromagnetic Energy. 2018;**52**(4):276-286

[17] Metaxas AC, Meredith RJ. Industrial Microwave Heating. London: The Institution of Engineering and Technology; 2008

[18] Macedo LL, Corrêa JL, Júnior IP, da Silva AC, Vimercati WC. Intermittent microwave drying and heated air drying of fresh and isomaltulose (Palatinose) impregnated strawberry. LWT. 2022; **155**:112918

[19] Szadzińska J, Mierzwa D. Intermittent–microwave and convective drying of parsley. In: IDS 2018. 21st International Drying Symposium Proceedings. València: Editorial Universitat Politècnica de; 2018. pp. 1455-1462

[20] Kalantari M. Microwave Technology in Freeze-Drying Process. In: Emerging Microwave Technologies in Industrial, Agricultural, Medical and Food Processing. London, UK: IntechOpen; 2018. pp. 143-157

[21] Wang W, Chen G. Freeze drying with dielectric-material-assisted microwave heating. AIChE Journal. 2007;**53**(12):3077-3088

[22] Duan X, Zhang M, Mujumdar AS. Studies on the microwave freeze drying technique and sterilization characteristics of cabbage. Drying Technology. 2007;**25**(10):1725-1731

[23] Duan X, Zhang M, Li X, Mujumdar AS. Ultrasonically enhanced osmotic pretreatment of sea cucumber prior to microwave freeze drying. Drying Technology. 2008;**26**(4):420-426

[24] Khoshtaghaza MH, Darvishi H, Minaei S. Effects of microwave-fluidized bed drying on quality, energy consumption and drying kinetics of soybean kernels. Journal of Food Science and Technology. 2015;**52**(8):4749-4760

[25] Goksu EI, Sumnu G, Esin A. Effect of microwave on fluidized bed drying of macaroni beads. Journal of Food Engineering. 2005;**66**(4):463-468

[26] Taheri S, Brodie G, Gupta D. Microwave fluidised bed drying of red lentil seeds: Drying kinetics and reduction of botrytis grey mold pathogen. Food and Bioproducts Processing. 2020;**119**:390-401. DOI: 10.1016/j.fbp.2019.11.001

[27] Monteiro RL, Carciofi BA, Marsaioli A Jr, Laurindo JB. How to make a microwave vacuum dryer with turntable. Journal of Food Engineering. 2015;**166**:276-284

[28] Bórquez R, Melo D, Saavedra C. Microwave–vacuum drying of strawberries with automatic temperature control. Food and Bioprocess Technology. 2015;**8**(2):266-276

[29] Sagar VR, Suresh KP. Recent advances in drying and dehydration of fruits and vegetables: A review. Journal of Food Science and Technology. 2010;**47**(1):15-26

[30] Calín-Sánchez Á, Szumny A, Figiel A, Jałoszy nski K, Adamski M, Carbonell-Barrachina Á. A effects of vacuum level and microwave power on rosemary volatile composition during vacuum-microwave drying. Journal of Food Engineering. 2011;**103**:219-227

[31] Wardhani NS, Amanda N, Sari AR. Microwave vacuum drying on fruit: A

review. In: 2nd International Conference on Smart and Innovative Agriculture (ICoSIA 2021). Paris: Atlantis Press International B.V.; 2022. pp. 309-316

[32] Dak M, Pareek NK. Effective moisture diffusivity of pomegranate arils undergoing microwave-vacuum drying. Journal of Food Engineering. 2014;**122**: 117-121

[33] Kumar V, Shrivastava SL. Vacuum-assisted microwave drying characteristics of green bell pepper. International Journal of Food Studies. 2017;**6**:67-81

[34] Zielinska M, Zielinska D, Markowski M. The effect of microwave-vacuum pretreatment on the drying kinetics, color and the content of bioactive compounds in osmo-microwave-vacuum dried cranberries (Vaccinium macrocarpon). Food and Bioprocess Technology. 2018;**11**(3): 585-602

[35] Berteli MN, Rodier E, Marsaioli A Jr. Study of the microwave vacuum drying process for a granulated product. Brazilian Journal of Chemical Engineering. 2009;**26**:317-329

[36] Chou SK, Chua KJ. New hybrid drying technologies for heat sensitive foodstuffs. Trends in Food Science & Technology. 2001;**12**(10):359-369

Chapter 5

Effects of Pretreatments with Ethanol and Ultrasound on Convective Drying of BRS Vitória Grapes

Nathalia Barbosa da Silva, Patrícia Moreira Azoubel and Maria Inês Sucupira Maciel

Abstract

The objective of this study was to evaluate the effect of ethanol and ultrasound as pretreatment to improve the convective drying of the BRS Vitória grape. The drying kinetics, rehydration, quality parameters, and phenolic compounds were evaluated. Before drying, grapes cv. BRS Vitória was ultrasound treated using two separate means, with ethanol (99.5% v/v) and distilled water. After pretreatment, the grapes were dried at 60°C and 0.1 m/s. The Logarithmic model provided a better prediction to describe the drying of grapes. Peleg's model showed satisfactory adjustments to predict rehydration. Compared to the Control, pretreatment using the combination of ultrasound and ethanol decreased the drying time of the grapes by 61%. The pretreatments did not influence in quality parameters. In contrast, phenolic retention was observed in samples with ethanol. These results open new perspectives on the drying process and product quality by combining ethanol and ultrasound.

Keywords: ultrasound, dehydration, ethanol pretreatment, raisin, logarithmic model

1. Introduction

Grape is a berry belonging to the Vitaceae family and is widely cultivated and frequently consumed in the world. According to the Food and Agriculture Organization (FAO), its world production in 2020 was approximately 100 million tons. The principal producers are China, Italy, Spain, and France. Currently, Brazil occupies the 15th position of grape producers, with a production of 1,435,596.00 tons in 2020 [1].

In Brazil, grapes are consumed in fresh or processed form as juices, vines, jams, and raisins. Part of the production of grapes in Brazil comes from the São Francisco valley, a region in northeastern Brazil with productive potential for different grape cultivars [2]. Therefore, Research Institutions have been developing grape cultivars adapted to Brazilian conditions to meet the high demand of the foreign market [3].

Grape cv. BRS Vitória was developed by a Brazilian agricultural research company (EMBRAPA) in 2012 to increase the production and improve climate adaptation of grapes in the country. Seedless grape, the productivity of this cultivar can exceed 30 t/ha and shows good tolerance to berry splitting and downy mildew. The berry is spherical, black in color, with thick and resistant skin and colorless pulp. This fruit could provide health-related benefits (rich in phenolics, anthocyanins, and flavonoids with antioxidant properties) [4].

However, the grapes have a high moisture and sugar content, reducing the shelf life of the fruit [5]. Drying is one of the most used conservation methods to increase the shelf life of perishable foods such as grapes. Drying reduces the food moisture content to a level that allows safe storage for an extended period, reducing weight and volume, and minimizing packaging, storage, and transport costs [6].

For food drying to occur effectively, it is necessary to evaluate the following issues: the drying kinetics and factors that affect the drying rate; product quality, since water removal is not the only consequence of the process. Other important quality-related changes in taste, flavor, appearance, texture, structure, and nutritive value may occur in the course of drying [7].

The intrinsic characteristics of the berries also influenced the drying process. Grapes have waxy skin, which makes it difficult to mass transfer [8, 9]. To remove the waxy layer and accelerate the dehydration process of the grapes, several pretreatments have already been applied and investigated, such as blanching, the alkaline emulsion of ethyl oleate solution (AEEO), abrasion, and carbonic maceration [10–12].

Some novel non-thermal technology like ultrasound has been employed to enhance the drying process. This technology could be used with pretreatment for their benefit in enhancing heat and mass transfer in the course of dehydration [13–15]. Ultrasonic waves cause structural changes in the products, enabling increased permeability of the material. This effect can be obtained due to the "sponge effect," cavitation phenomenon, and the effects accompanying cavitation, such as the formation of microchannels, facilitating mass, and/or heat transfer [16]. Ultrasound applications allow reducing drying time and energy consumption, obtaining high-quality dried materials [17]. This technology has been applied as a pretreatment in the drying of sweet potatoes [18], bitter melon [19], and kiwifruits [20]. The studies revealed that ultrasound pretreatment was effective to improve the process.

There are different types of immersion mediums used in ultrasound. Ethanol is an organic solvent with lower surface tension than water and facilitates the solvent into the food. Ren et al. [21] investigated the effects of different pretreatment methods on the drying process and the quality of catalytic infrared dried ginger slices. They observed sample pretreatment by ethanol + US had the highest drying efficiency and highest bioactive content retention. However, no studies have examined the effect of ultrasound combination as pretreatment on drying kinetics, quality parameters, and phenolic compounds from grapes.

Thus, the objective of this study was to evaluate the application of ultrasound as pretreatment to improve the convective drying of BRS Vitória grapes. For this purpose, the effect of an aqueous medium (ethanol and water) on drying kinetics, quality parameters, and phenolic compounds of raisins have been studied.

2. Materials and methods

2.1 Materials

For this study, grape cv. BRS Vitória was produced in the São Francisco Valley region (Latitude 09° 09 'South; Longitude 40° 22' West). The grapes were washed to remove surface impurities and sanitized with sodium hypochlorite (200 ppm) for 10 minutes. Then they were dried with absorbent paper, packed in polyethylene bags, and stored at −18 ± 1°C, until use.

2.2 Pretreatments

The pretreatments were conducted to evaluate the effect of ultrasound with different solvents (ethanol and distilled water). The sample (100 g) was placed in a beaker containing 200 mL of ethanol (99.5% v/v) encoded as US+ETOH. This beaker was then positioned in a thermostatic bath to maintain −5°C during sonication. The same process was conducted with 200 mL of distilled water and encoded as US+WATER. An ultrasonic probe (QR1000 Ultronique, Ecosonics, Brazil) with a constant frequency of 20 kHz, maximum power of 550 W, and microdot with a diameter of 25.4 mm was used. The operating time on the ultrasound was 30 minutes.

2.3 Drying process

Drying was performed for pretreated and untreated (control) grapes at 60°C ~ 1 m/s. For each batch, 100 g of grapes were placed on a metal net in a drying oven (MA035, Marconi, Brazil), with air circulation and renewal. All drying processes were performed by periodic weight (every 1 h). The initial moisture content was determined according to AOAC [22]. All experiments were repeated three times at the respective temperature, and the average measurements are contained within this study.

2.4 Mathematical modeling

To calculate the moisture content (MR) of the grape, the following equation was used (1).

$$MR = \frac{Mt - Me}{M_0 - Me} \tag{1}$$

Where Mt., M0, and Me are the moisture content at a given drying time (g water/100 g dry matter), the initial moisture content (g water/100 g dry matter), and the balance moisture content (g water/100 g dry matter), respectively. The drying curves were fitted with eight distinct thin layer dehydration equations to test which system most accurately described the drying process. These equations are listed in **Table 1**.

Various statistical parameters including the Coefficient of Determination (R2) and root mean square error (RMSE) were used to describe the best fit. The higher value of R2 and the lower value of RMSE indicate the goodness of fit. R2 and RMSE equations can be described by eqs. (2) and (3).

$$R^? = 1 - \frac{\sum_{i=1}^{N}\left(MR_{pre,i} - MR_{\exp,i}\right)^2}{\sum_{i=1}^{N}\left(MR_{\exp,i} - \overline{MR_{\exp}}\right)^2} \tag{2}$$

$$RMSE = \sqrt{\frac{1}{N}\sum_{i=1}^{N}\left(MR_{pre,i} - MR_{\exp,i}\right)^2} \tag{3}$$

Models	Equations	References
Newton	MR = exp.(−k t)	[23]
Page	MR = exp.(−k t n)	[24]
Henderson and Pabis	MR = a exp.(−k t)	[25]
Logarithmic	MR = a exp.(−k t) + c	[26]

Table 1.
Mathematical models provided by several authors for drying curves.

Where N and z are the number of experimental data values and the number of constants, respectively. MRexp,i, MRpre,i, and MRexp are the experimental moisture ratio, predicted moisture ratio at time t, and the mean of experimental moisture ratio, respectively.

2.5 Quality parameters

The quality parameters were evaluated in fresh and processed samples. For quality analyses, grapes were dried until a final moisture content of 20% (wet basis), which is a value within the range allowed by Brazilian legislation [27]. Water activity (aw) was determined in three repetitions for every sample (fresh and dried grapes) at a temperature 25°C, using equipment Aqualab 4TE (Meter group, USA) according to the manufacturer's instructions. One sample of the tested material was placed into the chamber of the apparatus and closed. After about 5 min, the results were determined [28]. The soluble solids data were obtained using a digital refractometer (r2 i300, Reichert, USA). Juice from the sample was extracted and inserted into the equipment for reading, and the results were expressed in °Brix [29]. All measurements were carried out in triplicate.

2.5.1 Texture

The texture was evaluated using a texture meter (CT3–1000, Brookfield, USA), with the aid of data acquisition software of the same equipment brand. The hardness of fresh grapes was evaluated according to the methodology described by Rolle et al. [30]. For raisins, the method described by Wang et al. [31] with some modifications was used. Compression tests were carried out by compressing the raisin to 5 mm on the mid-axis with a cylindrical probe of 25.4 mm in diameter, with a waiting time of 5 seconds between the two bites, and at a speed rate of 1 $mm.s^{-1}$ to determine hardness.

2.5.2 Color

The color parameters of grapes were determined by using a colorimeter (CR-400, Konica Minolta Sensing, Japan). The samples were analyzed and expressed as color coordinates in the CIELAB space where L^* (brightness–darkness), a^* ($+a^*$: red, $-a^*$: green), and b^* ($+b^*$: yellow, $-b^*$: blue). White tile was used as a standard (Y = 93,40; x = 0,3136; y = 0,3196). The parameters L (Luminosity), a^*, and b^* allowed the calculation of the Hue angle, that is, the color tone using the following eq. (4) [32]:

$$h = \tan^{-1}\left(\frac{b^*}{a^*}\right) \tag{4}$$

2.6 Total phenolic content

The phenolic compounds were extracted using an ultrasonic bath (USC-2850A, Unique, Brazil) and as a solvent, ethanol (60% acidified with 0.1% HCL). The total phenolics content present in this extract was quantified according to the methodology proposed by Wettasinghe & Shahidi [33] using Folin-Ciocalteu reagent and gallic acid as a reference standard. 0.5 mL of the extract was homogenized with 8 mL of distilled water, 0.5 mL of Folin Ciocalteau reagent, and 1 mL of saturated sodium carbonate solution. The flasks were shaken and then kept at rest, in the dark, for 1 h. The absorbance at 765 nm was measured using a UV-vis spectrophotometer (UV-1900i, Shimadzu, Japan), and the results were expressed in mg of total phenolics in gallic acid equivalent (EAG) per 100 g of fresh grape and 100 g of raisin of dry matter.

2.7 Statistical analysis

Nonlinear regression was used to find model parameters to fit drying kinetics data. For this, Origin Pro 2019b software (Origin lab Inc., USA) was used. All determinations were performed in triplicate, and the data were submitted to the two-way Analysis of Variance (ANOVA) and Tukey post hoc test at a 5% significance level using Statistica 10.0 software (StatSoft Inc., USA).

3. Results and discussion

3.1 Drying kinetics

Fresh grape samples used in this work presented a moisture content of 84.33 ± 0.9% (w.b), which was in the range (80.04 ± 1.10–84.01 ± 1.6%) reported by Okzan et al. [34] and Adietta et al. [5] for black "Isabel" and Red Globe grapes, respectively. Before starting drying, the ultrasound with different mediums (water and ethanol) was applied. The effect of each pretreatment on the processing time was compared with the control treatment, as shown in **Figure 1**.

Grapes have high moisture and require a long drying time. To reach equilibrium moisture, the time required for the control sample under drying conditions (60°C and 1 m/s) was 41 hours. **Figure 1** shows there was an effect on drying time reduction, indicating that ethanol was the medium that reduced the time by 61%, with a processing time of 16 hours, while the medium with water reduced the drying time by 17% (34 hours).

Rojas, Silveira and Augusto [35] studied the application of ethanol and ultrasound combined as pretreatment in the drying kinetics of pumpkin using air at 50°C. The authors observed that the combination of ethanol and ultrasound for 30 minutes reduced the drying time of pumpkin by 59% compared to the control. Da Cunha et al. [36] evaluated the effectiveness of the use of ethanol, ultrasound, and/or vacuum as a pretreatment to melon drying. They observed a reduction of 44.62% in drying time with the use of ultrasound associated with ethanol. The authors reported a positive effect on the drying rate with the combination of medium and ultrasound, similar to the results found in this study.

The moisture kinetics of grape cv. BRS Vitória under different treatments is illustrated in **Figure 2**. The moisture ratio with time showed an exponentially decreasing trend in all treatments. **Figure 2**, it was observed similar behavior on the drying curve

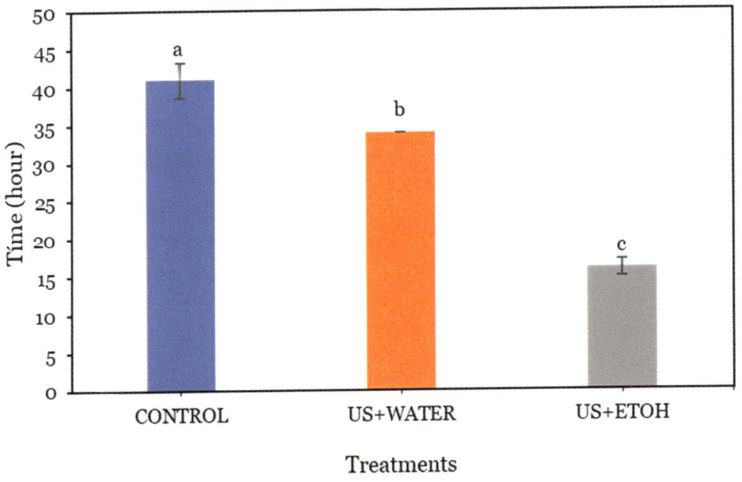

Figure 1.
Drying time reduction for all treatments. The values presented refer to the arithmetic mean of three determinations ± standard deviation. Equal letters do not differ statistically from each other at a 5% probability level by the Tukey test (ANOVA p < 0.05).

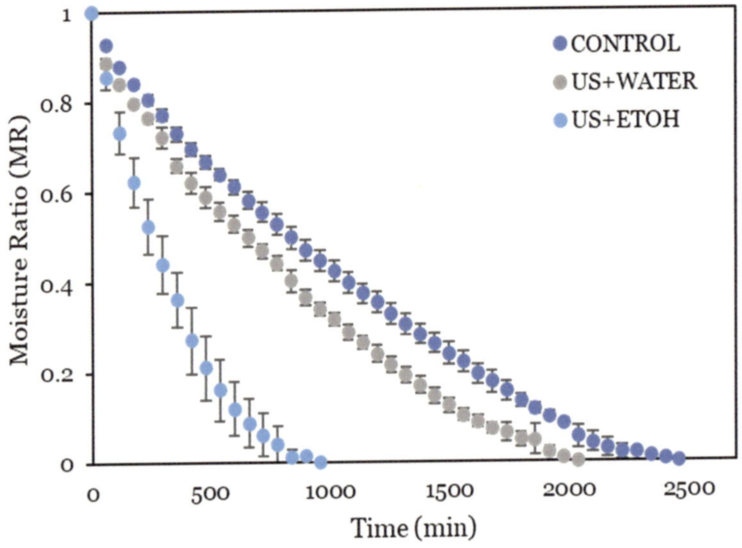

Figure 2.
Convective drying kinetics for all treatments. The values presented refer to the arithmetic mean of three determinations ± standard deviation.

for control and US+WATER samples. However, the longest drying time was obtained for the control sample. Drying kinetics is an important task to observe the behavior of the product during drying. The use of mathematical models is useful to design drying systems and analyze the complex phenomena of heat and mass transfer [37]. **Table 2** shows the statistical parameters estimated for the comparison between the four mathematical models of drying.

The best mathematical model was selected based on a comparison of the statistical values of the coefficient of determination (R2) and root mean square error (RMSE).

Treatment	Models*	Constants				R^2	RMSE
		k	*n*	*a*	*c*		
Control	Newton	0.0014	—	—	—	0.966	0.049
	Page	0.0015	0.9862	—	—	0.966	0.049
	Henderson e Pabis	0.0013	—	0.9447	—	0.971	0.045
	Logarithmic	0.0007	—	1.1526	−0.2671	0.991	0.025
US+ETOH	Newton	0.0033	—	—	—	0.986	0.035
	Page	0.0009	1.2093	—	—	0.996	0.020
	Henderson e Pabis	0.0034	—	1.0400	—	0.988	0.033
	Logarithmic	0.0028	—	1.0828	−0.0685	0.995	0.021
US+WATER	Newton	0.0012	—	—	—	0.962	0.057
	Page	0.0003	1.1965	—	—	0.981	0.039
	Henderson e Pabis	0.0012	—	1.0533	—	0.971	0.053
	Logarithmic	0.0004	—	1.7728	−0.7994	0.999	0.008

** All models were significant $p < 0.05$.*

Table 2.
Estimated parameters, coefficient of determination (R2) and root mean square error (RMSE), for mathematical models with and without ultrasound pretreatment.

The models fitted to the experimental data presented R2 values between 0.962 and 0.999 and the RMSE values were between 0.008 and 0.057, indicating that a good fit was obtained for all the proposed models. The logarithmic model presented the best fit for the drying processes performed in different treatments, indicating that in this model, changes in the moisture content of the grapes could be predicted with the drying time. The values of the constant k of the Logarithmic model indicated that with the decrease in the drying time, the constant increases. This behavior was observed with the pretreatment with ultrasound-assisted and ethanol medium.

3.2 Quality parameters

The results of the soluble solids content of fresh grapes and raisins in different treatments shown in **Figure 3**. In fruit drying, with the removal of moisture, the food content is concentrated and increases in the soluble solids content [38]. The soluble solids of BRS Vitória grapes dried with different treatments increased significantly than fresh grapes ($p < 0.05$). However, there was no difference between the treatments used and the control sample ($p > 0.05$). This result indicates that the media used do not affect the soluble solids content.

Water activity is an intrinsic factor in the food and indicates the free water contained in the food. This parameter is relevant to assess the stability of the product after processing [39]. Water activity below 0.6 can prevent the growth of microorganisms, increasing the shelf-life of dehydrated products during storage [40]. **Figure 4** compares the water activity of different treatments and fresh grapes. The water activity content for fresh grapes was 0.96. The treated samples ranged from 0.55 to 0.59 after drying. All dehydrated samples obtained water activity results below 0.6, guaranteeing the stability of the raisin. No significant differences were found between samples treated with different mediums and control samples. Similar behavior occurred in the soluble solids content.

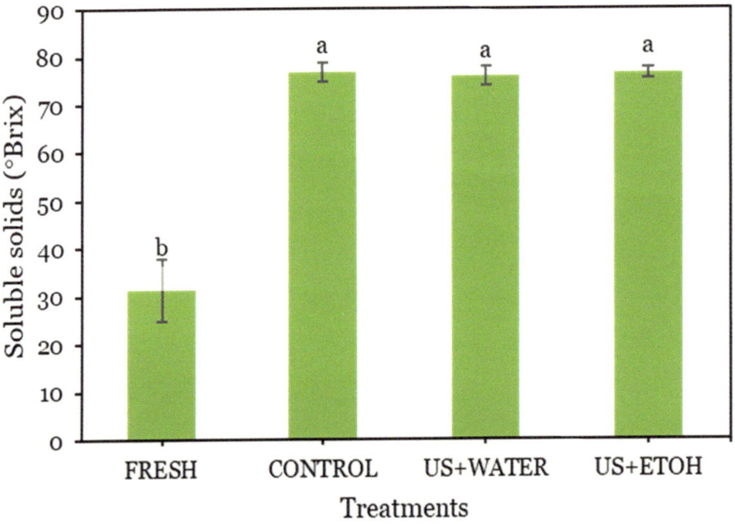

Figure 3.
Soluble solids content of BRS Vitória grapes in different treatments. The values presented refer to the arithmetic mean of three determinations ± standard deviation. Equal letters do not differ statistically from each other at a 5% probability level by the Tukey test (ANOVA p < 0.05).

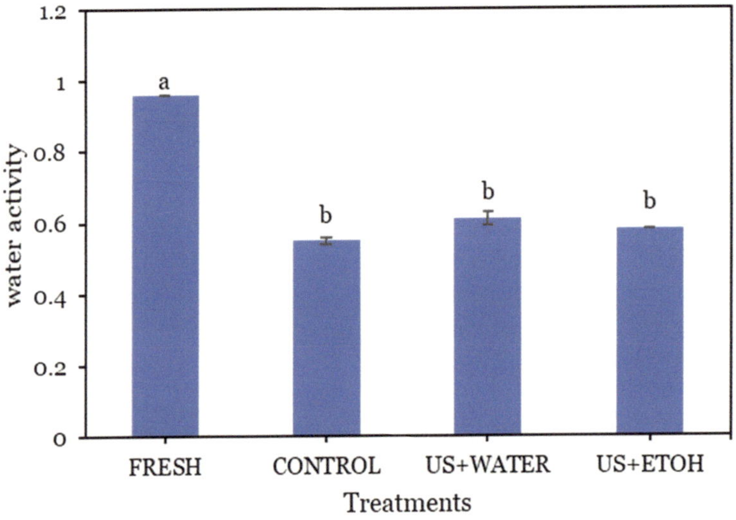

Figure 4.
Water activity of BRS Vitória grapes in different treatments. The values presented refer to the arithmetic mean of three determinations ± standard deviation. Equal letters do not differ statistically from each other at a 5% probability level by the Tukey test (ANOVA p < 0.05).

The instrumental color is one of the most important parameters to analyze the drying process. Color is measured using the L*a*b* system, in which L* indicates lightness, a* indicates color from green (−a*) to red (a*), and b* indicates color from blue (−b*) to yellow (b*). The changes in the values of the color parameters, mainly in the* and b* coordinates, it is possible to predict pigmentation changes or the occurrence of enzymatic or non-enzymatic browning reactions [37].

The results of the color parameters are shown in **Table 3**. The luminosity value (L*) of all samples decreased with drying. This result indicates that the raisins became

	Fresh	Control	US+WATER	US+ETOH
L*	2.29 ± 0.84	19.57 ± 2.39	21.42 ± 18.54	18.50 ± 1.59
a*	- 0.66 ± 0.17^{b}	1.58 ± 1.28^{a}	1.05 ± 0.20^{a}	1.37 ± 0.08^{a}
b*	1.61 ± 0.08	1.54 ± 0.15	1.44 ± 0.40	1.42 ± 0.32
Hue	112.00 ± 5.11^{a}	42.23 ± 26.23^{b}	53.09 ± 1.62^{b}	45.52 ± 7.33^{b}

***ANOVA p value<0.05. Means on lines followed by the same letters do not differ statistically from each other at the 5% probability level by the Tukey test.*
****values without letters were not significant p > 0.05.*The values presented refer to the arithmetic mean of three determinations ± standard deviation.*

Table 3.
Color parameters of BRS Vitória grape in different treatments.

opaquer. However, this coordinate showed no statistical difference. The values of a* coordinate for the control, ethanol, and water samples increased compared to fresh grapes, but no significant difference between the treatments ($p > 0.05$). There were no significant changes in the b* coordinate. The values obtained of hue angle for fresh grapes differed statistically from raisins ($p < 0.05$), showing a change in hue as an effect of drying. The results indicate that the drying of the grape causes changes in the luminosity, making it darker, with reddish and bluish nuances and with changes in tonality, regardless of the treatment used.

The hardness and chewiness of dried samples were evaluated by texture profile analysis TPA (**Figure 5**). The dried grapes presented values between 14.77 N and 31.65 N and the US+WATER treatment showed the highest value of hardness. There was a significant difference between the two treatments using ultrasound ($p < 0.05$). In the drying process, structural changes occur with the shrinkage of the product. The removal of moisture causes the surface of the sample to harden. Thus, the adhesive force between the cells forms a compact tissue when the water is removed [41]. It was observed that the raisin treated with ultrasound and ethanol is the one that necessitates less force for deformation.

In chewiness, raisins using ultrasound with ethanol had the highest average ($p < 0.05$), showing that the sample treated with ethanol needs more energy for the mastication forces. According to [42], the application of ultrasound pretreatment can cause significant changes in physical characteristics such as the hardness and chewiness of fresh food when subjected to drying. This behavior occurs due to the simultaneous transfer of heat and water during drying leading to tension and shrinkage, increasing the texture of the dehydrated products. However, in sonicated fruits, most of the cell walls are broken during ultrasonic vibration, and there is a network of micro-channels in the plant tissue, which favors the formation of a softer dried product.

3.3 Total phenolic content

The results of the total phenolic contents of BRS Vitória grape are presented in **Figure 6**. Total phenolic content was in the range of 340.98–1794.80 mg EAG/100 g. The TPC concentration of grape BRS Vitória increased with the drying process. Our results were in agreement with Serni et al. [43] determined TPC in dried grape pinot blanc skin during ripening in the range from 582.33 to 705.50 mg GAE/100 g and Ozakan et al. [34], who reported TPC for black Isabel grape of 351.89 ± 35.12 to

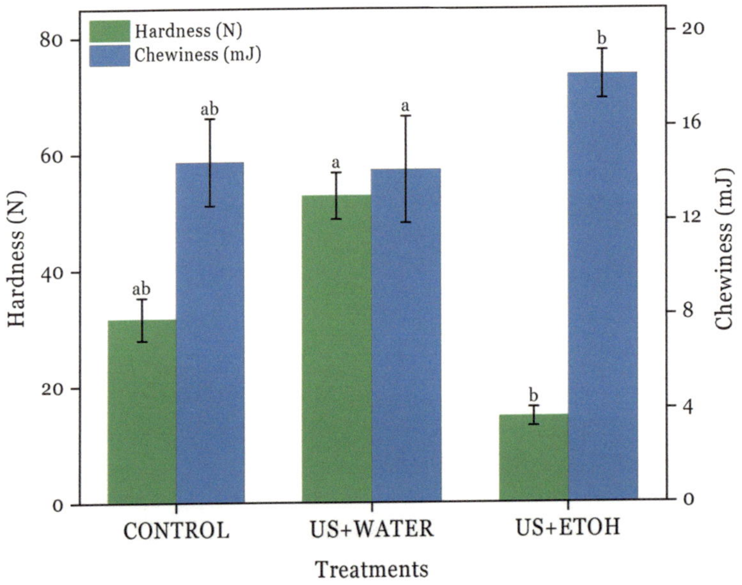

Figure 5.
Texture profile analysis (hardness and chewiness) of BRS Vitória grape in different treatments. The values presented refer to the arithmetic mean of three determinations ± standard deviation. Equal letters do not differ statistically from each other at a 5% probability level by the Tukey test (ANOVA p < 0.05).

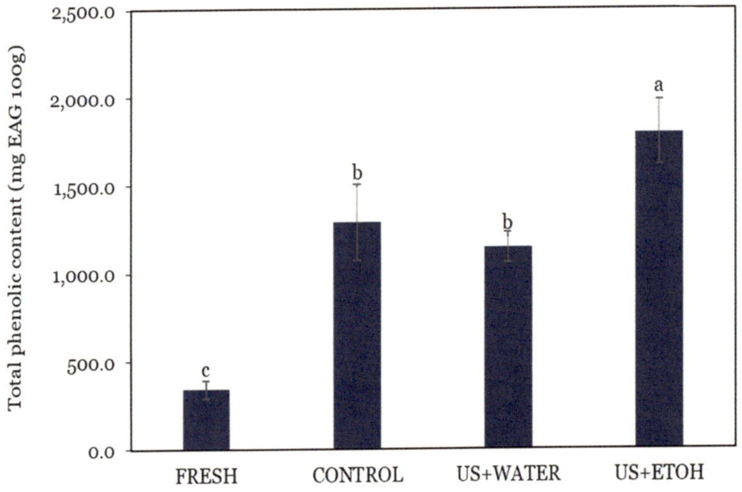

Figure 6.
Total phenolic content in BRS Vitória grape in different treatments. The values presented refer to the arithmetic mean of three determinations ± standard deviation. Equal letters do not differ statistically from each other at a 5% probability level by the Tukey test (ANOVA p < 0.05).

1101.61 ± 35.12 mg GAE/100 g. However, in this study, all drying methods reduced TPC concentration significantly. It should be noted that the US+ETOH treatment increased the TPC compared to the control and US+WATER samples ($p < 0.05$). It is due to the shortest drying time observed for US+ETOH treatment, as fewer phenolics were exposed to the heat, which increased the retention. Ren et al. [21] and Granella et al. [17] observed similar behavior for Chinese ginger and banana slices, respectively.

4. Conclusion

This work evaluated the effect of ultrasound with different media (water and ethanol) as a pretreatment in the convective drying of the BRS Vitória grape. The pretreatment with ultrasound in the different media increased the efficiency of convective drying of the BRS Vitória grape, reducing its drying time by up to 61% using ethanol. In addition, it was observed that, of all the mathematical models evaluated, the Logarithm was the best adjusted to the grape drying process when compared to the other models. In quality parameters of the raisin, no significant differences were observed between the media used and the control sample regarding texture, color, soluble solids, and water activity. Compared to fresh, no loss of phenolic content in grapes after drying. Ultrasound with ethanol combined showed the highest phenolic content between the treatments. Therefore, pretreatment with ethanol proved to be effective in obtaining raisins, reducing the drying time, not altering the quality characteristics of the product, and promoted more retention of nutrients.

Conflict of interest

The authors declare no conflict of interest.

Author details

Nathalia Barbosa da Silva[1], Patrícia Moreira Azoubel[2] and Maria Inês Sucupira Maciel[3]*

1 Technology Center, Federal University of Paraíba, João Pessoa, Brazil

2 Department of Chemical Engineering, Federal University of Pernambuco, Recife, Brazil

3 Department of Consumer Sciences, Rural Federal University of Pernambuco, Recife, Brazil

*Address all correspondence to: m.inesdcd@gmail.com

References

[1] Food and Agriculture Organization of the United Nations. FAOSTAT. 2020 http://www.fao.org/faostat/en/#data/QC

[2] Olivati C, de Oliveira Nishiyama YP, de Souza RT, et al. Effect of the pre-treatment and the drying process on the phenolic composition of raisins produced with a seedless Brazilian grape cultivar. Food Research International. 2019;**116**:190-199. DOI: 10.1016/j.foodres.2018.08.012

[3] Mapa MDAPE. Culturas: uva. 2015 http://www.agricultura.gov.br/vegetal/culturas/uva. [Accessed February 20, 2022]

[4] Maia JDG, Ritschel P, Camargo UA, et al. 'BRS Vitória' Nova cultivar de uva de mesa sem sementes com sabor especial e tolerante ao míldio. Bento Gonçalves: Embrapa; 2012

[5] Adiletta G, Russo P, Senadeera W, di Matteo M. Drying characteristics and quality of grape under physical pretreatment. Journal of Food Engineering. 2016;**172**:9-18. DOI: 10.1016/j.jfoodeng.2015.06.031

[6] Langová R, Jůzl M, Cwiková O, Kos I. Effect of different method of drying of five varieties grapes (Vitis vinifera L.) on the bunch stem on physicochemical, microbiological, and sensory quality. Food. 2020;**9**:1-15. DOI: 10.3390/foods9091183

[7] Berk Z. Dehydration. In: Berk Z, editor. Food Process Engineering and Technology. 3rd ed. Cambridge: Academic Press; 2018:513-566. DOI: 10.1016/B978-0-12-812018-7.00022-1

[8] Huang CC, Wu JSB, Wu JS, Ting Y. Effect of novel atmospheric-pressure jet pretreatment on the drying kinetics and quality of white grapes. Journal of the Science of Food and Agriculture. 2019;**99**:5102-5111. DOI: 10.1002/jsfa.9754

[9] Tao Y, Wang P, Wang Y, et al. Power ultrasound as a pretreatment to convective drying of mulberry (Morus alba L.) leaves: Impact on drying kinetics and selected quality properties. Ultrasonics Sonochemistry. 2016;**31**:310-318. DOI: 10.1016/j.ultsonch.2016.01.012

[10] Wang Y, Tao H, Yang J, et al. Effect of carbonic maceration on infrared drying kinetics and raisin qualities of red globe (Vitis vinifera L.): A new pre-treatment technology before drying. Innovative Food Science and Emerging Technologies. 2014;**26**:462-468. DOI: 10.1016/j.ifset.2014.09.001

[11] Kriaa K, Nassar AF. Comparative study of pretreatment on microwave drying of gala apples (Malus pumila): Effect of blanching, electric field and freezing. LWT. 2022;**165**:113693. DOI: 10.1016/J.LWT.2022.113693

[12] Zemni H, Sghaier A, Khiari R, et al. Physicochemical, phytochemical and mycological characteristics of Italia Muscat raisins obtained using different pre-treatments and drying techniques. Food and Bioprocess Technology. 2017;**10**:479-490. DOI: 10.1007/s11947-016-1837-4

[13] Rojas ML, Augusto PED, Cárcel JA. Ethanol pre-treatment to ultrasound-assisted convective drying of apple. Innovative Food Science and Emerging Technologies. 2020;**61**:1-12. DOI: 10.1016/j.ifset.2020.102328

[14] Miano AC, Rojas ML, Augusto PED. Combining ultrasound, vacuum

and/or ethanol as pretreatments to the convective drying of celery slices. Ultrasonics Sonochemistry. 2021;**79**:1-9. DOI: 10.1016/j.ultsonch.2021.105779

[15] Zhou C, Wang Z, Wang X, et al. Effects of tri-frequency ultrasound-ethanol pretreatment combined with infrared convection drying on the quality properties and drying characteristics of scallion stalk. Journal of the Science of Food and Agriculture. 2021;**101**:2809-2817. DOI: 10.1002/jsfa.10910

[16] Xu B, Sylvain Tiliwa E, Yan W, et al. Recent development in high quality drying of fruits and vegetables assisted by ultrasound: A review. Food Research International. 2022;**152**:110744. DOI: 10.1016/J.FOODRES.2021.110744

[17] Granella SJ, Bechlin TR, Christ D. Moisture diffusion by the fractional-time model in convective drying with ultrasound-ethanol pretreatment of banana slices. Innovative Food Science & Emerging Technologies. 2022;**76**:102933. DOI: 10.1016/J.IFSET.2022.102933

[18] Oladejo AO, Ma H, Qu W, et al. Effects of ultrasound on mass transfer kinetics, structure, carotenoid and vitamin C content of Osmodehydrated sweet potato (Ipomea Batatas). Food and Bioprocess Technology. 2017;**10**:1162-1172. DOI: 10.1007/s11947-017-1890-7

[19] Jin W, Zhang M, Shi W. Evaluation of ultrasound pretreatment and drying methods on selected quality attributes of bitter melon (*Momordica charantia* L.). Drying Technology. 2019;**37**:387-396. DOI: 10.1080/07373937.2018.1458735

[20] Roueita G, Hojjati M, Noshad M. Study of physicochemical properties of dried kiwifruits using the natural hypertonic solution in ultrasound-assisted osmotic dehydration as pretreatment. International Journal of Fruit Science. 2020;**20**:S491-S507. DOI: 10.1080/15538362.2020.1741057

[21] Ren M, Ren Z, Chen L, et al. Comparison of ultrasound and ethanol pretreatments before catalytic infrared drying on physicochemical properties, drying, and contamination of Chinese ginger (Zingiber officinale roscoe). Food Chemistry. 2022;**386**:132759. DOI: 10.1016/J.FOODCHEM.2022.132759

[22] AOAC, editor. Official Methods of Analysis of Association of Official Analytical Chemists. 20th ed. Washington: AOAC; 2016

[23] Vengaiah PC, Pandey JP. Dehydration kinetics of sweet pepper (Capsicum annum L.). Journal of Food Engineering. 2007;**81**:282-286. DOI: 10.1016/J.JFOODENG.2006.04.053

[24] Page GE. Factors Influencing the Maximum Rates of Air Drying Shelled Corn in Thin Layers [Dissertation]. West Lafayette: Purdue University; 1949

[25] Henderson SM, Pabis S. Grain drying theory: Temperature effect on drying coefficient. Journal of Agricultural Engineering Research. 1961;**6**:169-174

[26] Erbay Z, Icier F. A review of thin layer drying of foods: Theory, modeling, and experimental results. Critical Reviews in Food Science and Nutrition. 2010;**50**:441-464. DOI: 10.1080/10408390802437063

[27] Brasil. Resolução – RDC nº 12, de 2 de janeiro de 2001. Brasília: Regulamento Técnico sobre padrões microbiológicos para alimentos; 2021

[28] Macedo LL, Corrêa JLG, Petri Júnior I, et al. Intermittent microwave drying and heated air drying of fresh and isomaltulose (Palatinose) impregnated strawberry. LWT. 2022;**155**:112918. DOI: 10.1016/j.lwt.2021.112918

[29] Monteiro RL, Carciofi BAM, Laurindo JB. A microwave multi-flash drying process for producing crispy bananas. Journal of Food Engineering. 2016;**178**:1-11. DOI: 10.1016/j.jfoodeng.2015.12.024

[30] Rolle L, Torchio F, Giacosa S, Río Segade S. Berry density and size as factors related to the physicochemical characteristics of Muscat Hamburg table grapes (Vitis vinifera L.). Food Chemistry. 2015;**173**:105-113. DOI: 10.1016/J.FOODCHEM.2014.10.033

[31] Wang J, Mu W, Fang X, Mujumdar AS. Food and bioproducts processing pulsed vacuum drying of Thompson seedless grape: Effects of berry ripeness on physicochemical. Food and Bioproducts Processing. 2017;**106**:117-126. DOI: 10.1016/j.fbp.2017.09.003

[32] Noshad M, Ghasemi P. Influence of freezing pretreatments on kinetics of convective air-drying and quality of grapes. Food Bioscience. 2020;**38**:1-8. DOI: 10.1016/j.fbio.2020.100763

[33] Wettasinghe M, Shahidi F. Evening primrose meal: A source of natural antioxidants and scavenger of hydrogen peroxide and oxygen-derived free radicals. Journal of Agricultural and Food Chemistry. 1999;**47**:1801-1812. DOI: 10.1021/JF9810416

[34] Ozkan K, Karadag A, Sagdic O. The effects of different drying methods on the in vitro bioaccessibility of phenolics, antioxidant capacity, minerals and morphology of black 'Isabel' grape. LWT. 2022;**158**:113185. DOI: 10.1016/j.lwt.2022.113185

[35] Rojas ML, Silveira I, Augusto PED. Ultrasound and ethanol pre-treatments to improve convective drying: Drying, rehydration and carotenoid content of pumpkin. Food and Bioproducts Processing. 2020;**119**:20-30. DOI: 10.1016/j.fbp.2019.10.008

[36] da Cunha RMC, Brandão SCR, de Medeiros RAB, et al. Effect of ethanol pretreatment on melon convective drying. Food Chemistry. 2020;**333**:127502. DOI: 10.1016/j.foodchem.2020.127502

[37] Sahoo M, Titikshya S, Aradwad P, et al. Study of the drying behaviour and color kinetics of convective drying of yam (Dioscorea hispida) slices. Industrial Crops and Products. 2022;**176**:114258. DOI: 10.1016/j.indcrop.2021.114258

[38] Venkatram A, Padmavathamma AS, Rao BS. Influence of storage temperature on sugars, Total soluble solids and acidity of raisins prepared from seedless varieties of grape (Vitis vinifera L.). International Journal of Current Microbiology and Applied Sciences. 2017;**6**:2095-2102

[39] Wiktor A, Parniakov O, Toepfl S, et al. Sustainability and bioactive compound preservation in microwave and pulsed electric fields technology assisted drying. Innovative Food Science and Emerging Technologies. 2021;**67**:1-6. DOI: 10.1016/j.ifset.2020.102597

[40] Dadan M, Nowacka M. The assessment of the possibility of using ethanol and ultrasound to design the properties of dried carrot tissue. Applied Sciences. 2021;**11**:689. DOI: 10.3390/APP11020689

[41] Xu W, Islam MN, Cao X, et al. Effect of relative humidity on drying characteristics of microwave assisted hot air drying and qualities of dried finger citron slices. LWT. 2021;**137**:1-10. DOI: 10.1016/j.lwt.2020.110413

[42] da Silva V, Júnior E, Lins de Melo L, Batista de Medeiros RA, et al. Influence of ultrasound and vacuum assisted drying on papaya quality parameters. LWT. 2018;**97**:317-322. DOI: 10.1016/J.LWT.2018.07.017

[43] Serni E, Tomada S, Haas F, Robatscher P. Characterization of phenolic profile in dried grape skin of Vitis vinifera L. cv. Pinot blanc with UHPLC-MS/MS and its development during ripening. Journal of Food Composition and Analysis. 2022;**114**:104731. DOI: 10.1016/j.jfca.2022.104731

Chapter 6

Summary of Investigations in Regard to the Kinetics of Absorbed Water Dehydration from Different Hydrogels

Jelena D. Jovanović and Borivoj K. Adnadjević

Abstract

A review of novel kinetics models of dehydration (DH) of equilibrium swollen hydrogels: poly(acrylic acid) hydrogel (PAAH), poly(acrylic-co-methacrylic acid) (PAMAH), and poly(acrylic acid)-g-gelatin (PAAGH), is presented. Kinetic curves of isothermal and non-isothermal dehydration of hydrogels were measured using thermogravimetric methods. The kinetic complexity of the dehydration process was analyzed by different methods: integral, differential, Kissinger-Assakura-Sanura (KAS), and Vyzovkin's method. The complex kinetics of dehydration of hydrogels was described by a series of new kinetic models: distribution apparent energy activation model (DAEM), Webull's distribution of reaction times, the dependence of the degree of conversion (α) on the temperature which is defined by the logistic function, coupled single step-approximation and iso conversion curve. Procedures were developed for calculating the function of the density distribution of probability ($g(E_a)$) of apparent activation energy (E_a). The relationship between the phase state of the absorbed water in hydrogel and the form of function of distribution of apparent E_a and kinetic parameters of dehydration was analyzed.

Keywords: dehydration, hydrogel, kinetics, models, water

1. Introduction

The existence of hydrogels dates from the 1960s when Wichterle and Lim first hypothesized the possibility of using the cross-linked hydrophilic polymer poly(2-hydroxyethyl methacrylate) (PHEMA) for contact lenses [1]. Hydrogels are three-dimensional (3D) cross-linked structures that are mainly composed of hydrophilic homopolymers or copolymers connected by chemical or physical bonds, which have the ability to absorb a significant amount of water or other biological fluids, without dissolving or losing their structural integrity (swelling) [2]. They represent a unique class of macromolecular materials with specific physicochemical properties: the ability to retain a large amount of water solution in their structure (from 20% to several 10,000% in relation to the weight in the dry state), the ability to possess a high degree of flexibility similar to natural tissue and exhibit physical, chemical and mechanical stability in the swollen state [3]. The amount of water that these materials can absorb and thus increase their initial volume is fascinating and exceeds by several orders of magnitude the amount of aqueous solution that other gels can. They are also called "smart", "intelligent", "stimuli-responsive" or "environmental-sensitive" and attracted great attention in recent years [4].

Due to the large amount of water that they can absorb, hydrogels are often called superabsorbent materials. The hydrogels' ability to absorb huge amounts of water results from the presence of side hydrophilic functional groups on their polymer chains. On the contrary, their resistance to dissolution is a consequence of crosslinking between the polymers' chains. The water inside the hydrogel allows free diffusion of dissolved molecules, while the polymer acts as a matrix that holds the water [5].

Hydrogel in the dry state is called xerogel, and it is a solid and brittle material that shows the typical properties of solid substances that are the result of cross-linking. Hydrogels are called permanent or chemical gels when they are covalently crosslinked, or physical gels when entanglements, weak associations of the van der Waals type, or hydrogen bonds formed a network. Crosslinks between polymer chains create the structure of the hydrogel network and give them physical integrity [6, 7]. Hydrogels can be natural, synthetic, semi-synthetic, or their combinations. Hydrogels can give a response i.e. react to external stimuli (temperature, nature of the solvent, pH value, ionic strength, electric or magnetic field action, light, biological agents, radiation, etc. through notable changes in their macroscopic properties, which are most frequently manifested through the changes in volume [8, 9]. Depending on the design of the polymer network, these volume changes can occur continuously or discontinuously over a certain range of stimulus changes or at a certain critical value of the stimulus.

Due to their specific properties, hydrogels become one of the upcoming classes of materials that have found wide applications in various fields. Among the numerous areas of application of hydrogels, the ones in the field of medicine and pharmacy are particularly significant [10, 11], especially in controlled and targeted drug release [12, 13], regenerative medicine and tissue engineering, contact lenses, biosensors, etc. [14]. Beyond their biomedical applications [15, 16], hydrogel represents an ideal basis for new materials for applications in biotechnology, environmental protection, agriculture, agrochemistry, horticulture, cosmetics, as superabsorbent in hygiene products, packaging materials for storing food, textile materials for special purposes, in sensor materials, etc. In recent times, the applications of hydrogels and hydrogel-derived materials (HDM) presents emerging novel materials platform in electrochemical energy conversion systems, including metal-air batteries, fuel cells, and water-splitting electrolyzers, due to their specific and tailorable physicochemical properties [17].

The broadest functional applications of hydrogels are founded on their distinguishing capability to reversibly absorb (swell) and release (dehydrate) water. Exactly because of that, and since water removal and uptake includes fundamental principles of physics, knowledge of the mechanism and kinetics of both swelling and dehydration of hydrogels is of the utmost importance, in order to be able to optimize their efficiency in particular applications. Despite the fact that the phenomenon of swelling of hydrogels, including its mechanism and kinetics, is among the most studied from a fundamental aspect in the field of hydrogels [18, 19], on the contrary, however, the same cannot be said for the dehydration of hydrogels. Hydrogel dehydration itself has been much less studied, since the analysis of the dehydration process in hydrogel materials is an extremely difficult task requiring complex approaches [20], and in particular there is a very limited number of published papers in the literature that deal with the issue of the mechanism and kinetics of dehydration.

Since the discovery and first application of hydrogels are related to contact lenses, and the state of hydration, that is dehydration, is very important for this application, there are a number of works dealing with the dehydration of contact lenses. However, on the one hand, they mostly relate to bio-medical and physiological aspects, [21–23] and they primarily relate to silicone lenses [24], while physicochemical studies of the kinetics of dehydration are much rarer.

In vitro dehydration of new and worn contact lenses by using different types of hydrogels including one silicone-hydrogel was investigated. It was found that contact lenses based on hydrogels exhibited different dehydration behavior, the different behavior during dehydration that was manifested corresponds to different rates of dehydration and stages of dehydration [21].

The dehydration process in hydrogels used in ophthalmology as intraocular lenses were investigated by Chamerski et al., using equilibrium-swelled hydrogels in deionized water and saline solution. Studies of the dehydration process were carried out by use of gravimetric analysis, Fourier-Transform Infrared, and Positron Annihilation Lifetime Spectroscopy. Obtained results revealed changes in hydrogen bonding structure and free volume holes induced by saline solution ingredients. Observation of the process at the molecular level has given information about water transport in the free volume holes on the basis of changes in hydrogen bonds and demonstrated a more filled and hydrogen-bonded structure in the case of fluid containing inorganic compounds. More stable network formation can be explained by the influence of such compounds on changes in water binding, and thus in internal structure transformation toward its improvement [20].

The dehydration kinetics of the poly(vinyl alcohol) (PVA) hydrogel aimed at wound dressing materials has been investigated. The effects of the thickness and initial water content amount on the dehydration process were evaluated. The results showed that the dehydration rate of the PVA hydrogel wound dressing has an inverse dependency on the hydrogel's thickness while the initial water content has no significant effect. The authors developed a mathematical model on the basis of the diffusion mechanism to predict the dehydration process of the wound dressings and the obtained results confirmed that the main phenomenon governing the dehydration of the wound dressings is diffusion [25].

Hawlader et al. [26] used the one-dimensional diffusion model to describe the transfer of heat and mass from the wet to the dry region of the hydrogel during dehydration. Water diffusion during the dehydration of polyacrylamide (PAAm) hydrogel was investigated by Roques et al. [27]. Based on the obtained results, they suggested a mathematical model that was able to well describe the diffusion kinetics of water during hydrogel dehydration. Kept et al. examined the applicability of different kinetic models for mathematically describing the kinetics of hydrogel dehydration/ drying [28]. The research group of Peckan developed a fluorescence technique for *in situ* monitoring gelation, swelling, and dehydration processes of various hydrogel based on determining changes of fluorescent spectra of gels formed by solution-free copolymerization, [29], dehydration of κ-carrageenan gels at different temperatures [30], and dehydration of PAAm hydrogels with various cross-linking degrees [31].

The kinetics of non-isothermal dehydration (NIT) of polyacrylic acid hydrogels (PAAH) has been investigated using various kinetic methods such as Kissinger,

Coats-Redfern, Van-Krevelen, and Horowitz-Metzger [32, 33]. Kinetic of non-isothermal dehydration of a silver nanocomposite hydrogel of poly(acrylic acid) grafted onto salep which was not possible to describe the complete dehydration process by a single mechanism have been investigated [34].

Dehydration of chitosan fibers-enhanced gellan gum hydrogel and chitosan fibers-enhanced polysaccharide hydrogels have been investigated and established two distinct kinetics stages: diffusion and nucleation [35, 36]. Ma et al. investigated dehydration kinetics of poly(vinyl alcohol)/poly(vinyl pyrrolidone)/hydroxyapatite composite hydrogel and found that consists of water diffusion through hydrogel network and evaporation [37].

Non-isothermal dehydration of equilibrium swollen PAAH [38, 39], poly(acrylic-co-methacrylic acid) (PAMAH) [38], and poly(acrylic acid)-g-gelatin (PAAGH) [33] hydrogel has been investigated. The kinetics of isothermal dehydration of equilibrium swollen PAAH [40] and PAMAH [41] was presented. The comparative kinetic study of NIT and isothermal dehydration (IT) of PAAH was performed [42, 43], as well as on IT kinetics of water evaporation and PAAGH [44]. The fluctuating (changing) structure of hydrogels during dehydration has been found [42].

Belich et al. extensively investigated the dehydration of alginate hydrogels including various approaches. The effects of operational procedures and other parameters, calcium, and alginate concentration, and the addition of biopolymer co-solutes on water evaporation from alginate gel beads have been investigated [45]. The non-isothermal water evaporation for a series of alginate-based gel beads was performed aimed at understanding the state of water. The observed shoulders at high temperatures of thermogravimetric curves (TG) have been ascribed to evaporation of water molecules [45]. The investigations of water evaporation from alginate gel beads showed that calorimetric approach to hydrogel matrix release properties can be used as the predicting tool for the diffusion of solvents [46]. The quasi-isothermal dehydration of thin films of pure water and aqueous sugar solutions was investigated. The effect of sugar on the dehydration process was evaluated. It was established that the trehalose molecules slow down the diffusion of water molecules through the substrate [47].

Jovanovic and Adnadjevic et al. investigated NIT of Ca-alginate hydrogel for the first time. The dependence of apparent activation energy on the degree of dehydration was determined by Friedman's differential iso conversion method. It was shown for the first time that the kinetics of NIT of Ca-alginate hydrogel can be successfully described entirely by the statistical model of hydrogel dehydration. The existence of two-phase states of water absorbed on the Ca-alginate hydrogel was confirmed and related to the observed changes in the values of kinetic parameters, at a constant heating rate, with temperature [48].

The effects of e microwave irradiation on hydrogel DH were investigated by Adnadjevic and Jovanovic and co-workers. The isothermal kinetic of water evaporation and PAAGH dehydration were investigated under microwave heating conditions (MWH). The IT kinetic curves of water evaporation and hydrogel DH have been mathematically described complete by the Polanyi–Winger equation. The resonant transfer of a certain energy amount from the reaction system to the libration vibration of molecules of water is suggested as the mechanism of water molecules' activation, both for evaporation and dehydration [49].

The main goal of this chapter is to get a deeper insight into the essence of newly established kinetic models of hydrogel dehydration in order to expand knowledge about the mechanism and kinetics of hydrogel dehydration.

2. Experimental part

2.1 Hydrogel synthesis

Poly(acrylic acid) hydrogel (PAAH), which has been applied for this investigation was synthesized by a procedure based on the simultaneous radical polymerization of acrylic acid and cross-linking of the formed poly(acrylic acid), according to the general procedure described in details [50]. Poly(acrylic acid-co-methacrylic acid) hydrogel (PAMAH) was synthesized by a procedure of radical co-polymerization of acrylic acid and methacrylic acid (1:1 mol ratio) and cross-linking of the polymers formed, using the thoroughly described procedure [51]. Synthesis of poly(acrylic acid)-g-gelatin hydrogel (PAAGH) was described in detail [44].

2.2 Preparation of equilibrium swollen hydrogels

Synthesized hydrogels have been washed-out, as described in hydrogel synthesis procedures, and subsequently air-dried in laboratory oven under a defined temperature regime until constant mass. The obtained products (xerogels) were stored in a vacuum exicator until use. With the aim to evaluating dehydration kinetics, the xerogels samples were grounded and allowed to swell (24 h) in bidistilled water at ambient temperature to ensure to reach the equilibrium state. The equilibrium swollen samples were undertaken, and excess water was drained and wiped with tissue paper to remove surface water immediately before the dehydration experiment.

2.3 Thermogravimetric measurements

2.3.1 Non-isothermal thermogravimetric measurements

NIT curves were recorded by a Du Pont thermogravimetric analyzer TGA model 9510. The analyses were performed with 20–25 $\pm$ 1 mg samples of equilibrium swollen hydrogel in platinum pans under nitrogen atmosphere, N_2 purity 5.0, and gas flow rate of 10 mL min^{-1}. Samples were heated in the temperature range from ambient temperature to 500 K with different heating rates from 5 to 30 K min^{-1}.

2.3.2 Isothermal thermogravimetric measurements

The isothermal mass loss experiments were carried out using a TA Instruments-SDT simultaneous TGA-DSC thermal analyzer model 2960. The analysis was performed with 20 $\pm$ 1 mg samples of equilibrium swollen hydrogel in platinum pans under a nitrogen atmosphere at a gas flow rate of 10 ml min^{-1}. Isothermal runs were performed at nominal temperatures of 306 K, 324 K, 345 K, and 361 K. The samples were heated from the start to the selected dehydration temperature at a heating rate of 300 K min^{-1} and then held at that temperature for a given reaction time.

2.4 Calculation of the dehydration degree

The degree of dehydration was calculated as:

$$\alpha = \frac{m_0 - m}{m_o - m_t} \tag{1}$$

where m_0, m, m_f refers to the initial, actual, and final masses of the sample at a thermogravimetric curve.

2.5 Mathematical consideration

All the experimental data fitting has been performed by using Origin Program and Levenberg-Marquardt method. The coefficient of determination R^2 is used as an error function to minimize the error distribution between the experimental data and model.

3. Results and discussion

3.1 Kinetics of non-isothermal dehydration of equilibrium swollen PAAH

Kinetics of non-isothermal dehydration of equilibrium swollen PAAH have been investigated in detail in the work of Adnadjevic et al. [39]. **Figure 1** shows: (a) TG curves and (b) conversion curves ($\alpha = \alpha\,(T)_{vh}$) of NITD of PAAH experimentally obtained at different heating rates (v_h).

The conversion curves, at all of the investigated heating rates, are asymmetric by shape. The increase in the heating rates leads to the increase in the values of the inflection temperature (T_p), the final temperature (T_f), as well as in the degree of asymmetry.

Since the conversion curves exhibited complex shape, Friedman's differential iso-conversional method [52] was applied to determine the dependencies of $E_{a,\alpha}$ and ln $[A_\alpha f(\alpha)]$ from the degree of dehydration. The dependency $E_{a,\alpha}$ on α is shown in **Figure 2**, whereas **Figure 3** shows the dependency $\ln[A_\alpha f(\alpha)]$ on α.

The values of kinetics' parameters decrease in a complex manner with increasing α. The complex change in the value of the kinetic parameters with the α indicates that the NIT of poly(acrylic acid) hydrogel is a kinetically complex process. By analyzing the shape of dependence $E_{a,\alpha}$ and $\ln[A_\alpha f(\alpha)]$ on α, it can be concluded that there are two characteristic shapes of change on the dependence curves $E_{a,\alpha}$ and $\ln[A_\alpha f(\alpha)]$ on α. For $\alpha \leq 0.2$ the values of $E_{a,\alpha}$, and $\ln[A\alpha f(\alpha)]$ decrease abruptly and almost linearly with an increase in α, while for $\alpha \geq 0.2$ the increase in α leads to a slow decrease in the

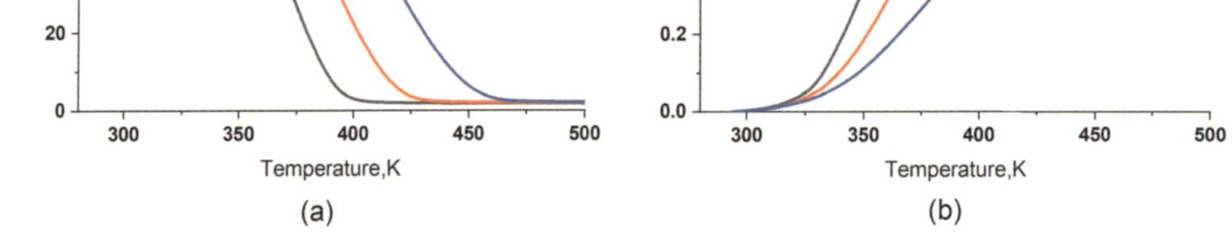

Figure 1.
TG curves (a) and conversion curves (b) of NIT of PAAH at different heating rates: (black line)—5°/min; (red line)—10°/min; (blue line)—20°/min.

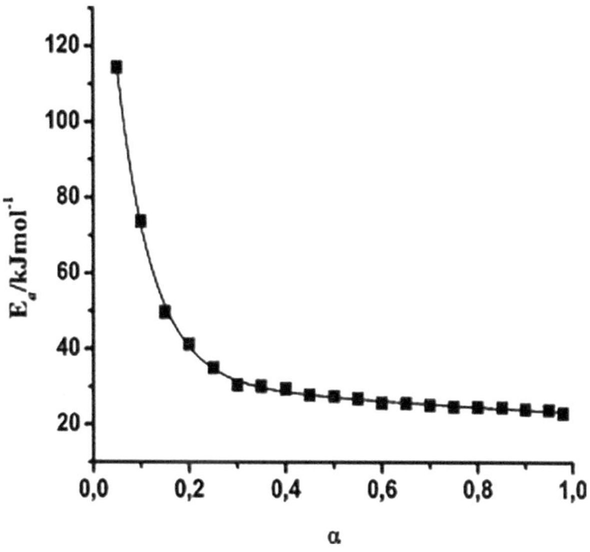

Figure 2.
The dependence of apparent E_a *on* α *[39].*

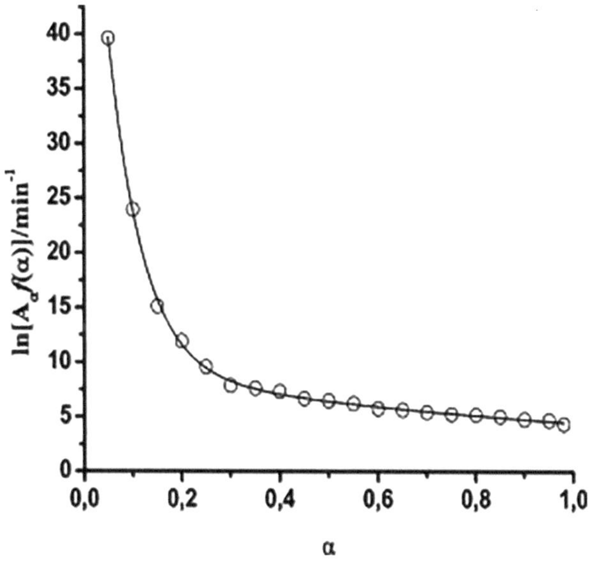

Figure 3.
The dependence of $ln[A_\alpha f(\alpha)]$ *on* α *[39].*

values of $E_{a,\alpha}$ and $\ln[A_\alpha f(\alpha)]$. Between the values of $\ln[A_\alpha f(\alpha)]$ and $E_{a,\alpha}$ there is a relationship, the so-called compensation effect defined by the relation:

$$ln\,[A_\alpha f(\alpha)] = -4.14 + 0.38E_{a,\alpha} \tag{2}$$

The existence of characteristic shape of changes in the value of kinetic parameters with α indicates that the phase state of the absorbed water in the hydrogel change with α, and that absorbed water exists in two different phase states. Most frequently for the mathematical description of complex chemical reactions, the so-called distributed activation energy model (DAEM) is used. DAEM is based on the assumption that a complex chemical reaction can be modeled as an infinite number of irreversible first-order paralleled reactions with the different values of Ea and A [53]. According to DAEM, the degree of dehydration can be calculated based on the eq. [54] (3):

$$\propto = 1 - \int_0^\infty exp\left[- \int_0^t Aexp\left(-\frac{Ea}{Rt} \right) dt \right] g(E_a) dE \tag{3}$$

where $g(E_a)$ is the density distribution probability of apparent activation energies. The density distribution probability of apparent activation energies is defined by the expression:

$$g(E_a) = \frac{d\alpha}{dE_a} \tag{4}$$

The expression (4) can be written in the form (5):

$$g(E_a) = \left(\frac{d\alpha}{dT} \right) \left(\frac{dT}{dE_a} \right) \tag{5}$$

using which, based on the knowledge of the dependence $\left(\frac{d\alpha}{dT}\right)_{v_h}$ and $\left(\frac{dT}{dE_a}\right)$ at a certain heating rate, $g(E_a)$ can be calculated at that heating rate. **Figure 4** shows the calculated values $g(E_a)$.

The density function of the probability distribution of apparent activation energies is in the shape of a narrow symmetrical peak with a maximum probability density, $g(E_a)_{max}$ = 0.067 at E_a = 23.9 kJ/mol. The shape of $g(E_a)$ is independent of the heating rate of the system. Therefore, the kinetic complexity of NITD of absorbed water on PAAH is a consequence of the distribution of reactivity of water molecules in the

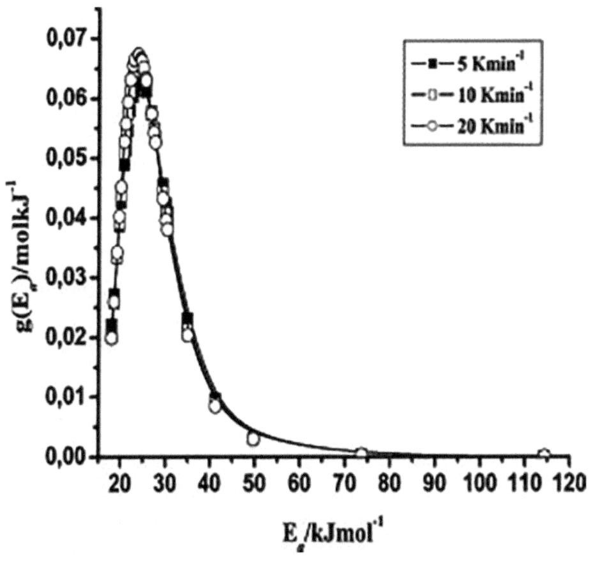

Figure 4.
The effect of v_h on the shape of $g(E_a)$ *[39].*

hydrogel, due to which there is a change in E_a and lnA with a change in the degree of dehydration.

3.2 Kinetics of isothermal dehydration of equilibrium swollen PAAH

Kinetics of IT of PAAH has been investigated in detail [40]. TG and conversion curves of IT of PAAH at different temperatures are shown in **Figure 5a** and **b**.

The conversion curves, at all of the investigated temperatures, have a similar asymmetric sigmoidal shape. With the increase in temperature, the degree of asymmetry of the curves increases, while the duration of dehydration shortens. In order to determine the degree of kinetic complexity of ITD using the integral iso conversion method [55] the shape of dependence of E_a, and $\ln[A/g(\alpha)]$ was determined. **Figure 6a** and **b** show the dependence of $E_{a,\alpha}$ and $\ln[A/g(\alpha)]$.

The values of kinetics parameters of isothermal dehydration of adsorbed water in PAAH increase complexly with an increase in the value of α. The complex change in values of kinetics parameters indicates that dehydration is a kinetically complex process.

Assuming that: (a) the degree of IT at a certain moment of time ($\alpha(t)_T$) is proportional to the probability of dehydration of the dehydrating centers at that moment of time ($p(t)_T$) = ($\alpha(t)_T$) Eq. (6); (b) the probability of dehydration of the dehydrating centers at that moment of time can be described by the Weibull function of distribution of the probability of the reaction times of the dehydrating centers [56], given with Eq. (6) at follows (7):

$$p(t)_T = 1 - exp\left[-\left(\frac{t}{\eta_T}\right)^{\beta_T}\right] \tag{6}$$

$$a(t)_T = 1 - exp\left[-\left(\frac{t}{\eta_T}\right)^{\beta_T}\right] \tag{7}$$

Figure 7 shows a comparison of experimental data (symbols) and calculated values of α (solid line) at T = 306 K and 361 K as an example.

Based on the results shown in **Figure** 7, it can be concluded that the conversion curves of IT absorbed water in PAAH can be completely described mathematically by Eq. (8)

$$a(t)_T = 1 - exp\left[-\left(\frac{t}{\eta_T}\right)^{\beta_T}\right] \tag{8}$$

Table 1 shows the effects of temperature on parameters β_T and η_T of the Weibull distribution function.

The values of the shape parameter increase with increasing temperature in accordance with Eq. (9):

$$\beta(T) = 10.85\,exp\left[-\left(\frac{635}{T}\right)\right] \tag{9}$$

In contrast, the values of the scale parameter η decrease with the increase in temperature in accordance with Eq. (10):

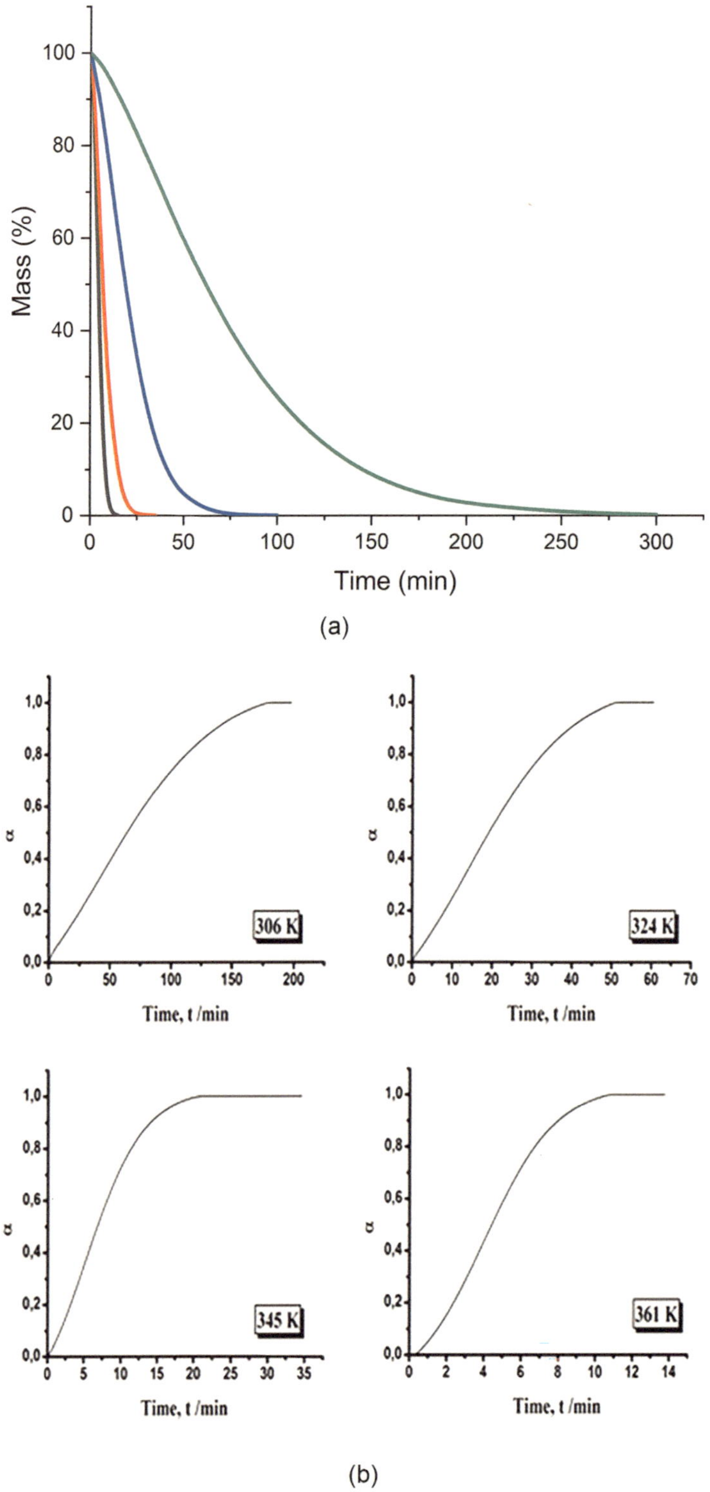

Figure 5.
(a) TG curves of ITD of PAAH at different temperatures (black line—306 K; (red line)—324 K; (blue line)—346 K, green line—361 K) and (b) conversion curves of ITD of PAAH at different temperatures [40].

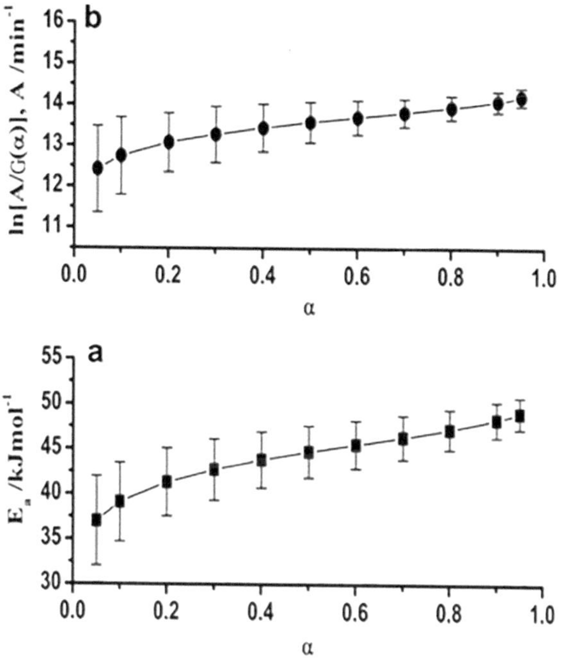

Figure 6.
Dependence of apparent E_a (a) and (b) $ln[A/g(\alpha)]$ on α.

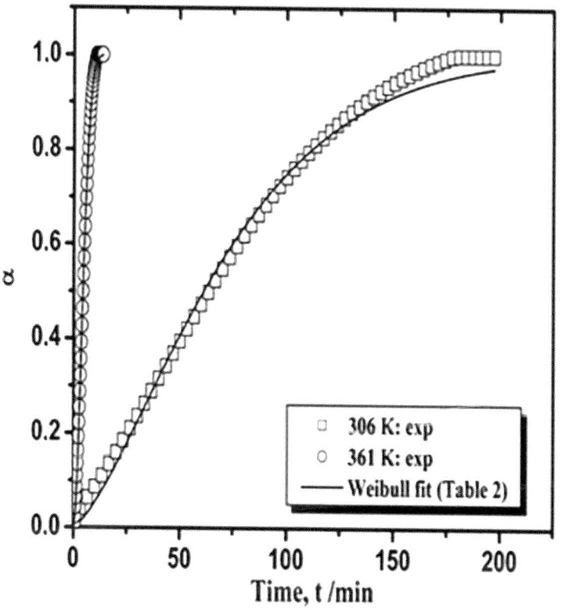

Figure 7.
Experimental conversion curves of PAAH at: (□) 306 K and (○) 361 K.

Temperature, (K)	β_T	η_T (min)	R^2
306	1.41	79.37	0.9960
324	1.49	23.42	0.9964
345	1.58	8.37	0.9989
361	1.94	5.23	0.9993

Table 1.
The effects of temperature on parameters β_T and η_T of the Weibull distribution function.

Temperature, T (K)	$t_{0,T}$ (min)	$\varepsilon_{0,T}$ (mol kJ^{-1})	R^2
306	0.00139	0.23925	0.9999
324	0.00074	0.22634	0.9999
345	0.00048	0.21340	0.9999
361	0.00184	0.17372	0.9999

Table 2.
Effect of temperature on the values of fitting coefficients $t_{o,T}$ and $\varepsilon_{o,T}$ for the ITD of PAAH.

$$\eta(T) = 3.70x10^8 \, exp\left[-\left(\frac{T}{20}\right)\right] \tag{10}$$

Complete description of isothermal conversion curves by Eq. (8) and knowledge of the functional dependence of t on $E_{a,\alpha}$ (Eq. (5)) enables the calculation of $g(E)$ (Eq. (11)).

$$g(E_a) = \left|\left(\frac{d\alpha}{dt}\right)\left(\frac{dt}{dE_a}\right)\right| \tag{11}$$

Based on the mathematical dependencies: $\alpha = f(t)_T$ the dependence $Ea,\alpha = Ea,\alpha$ $(\alpha)_T$ can be transformed into the dependence $E_{a,\alpha} = E_{a,\alpha}\,(t)_T$, i.e. the dependence $t = t$ $(E_{a,\alpha})_T$. Numerical analysis revealed that the $t = (E_{a,\alpha})_T$ can be mathematically described by Eq. (12).

$$t = t_{o,T}\, exp\left(\varepsilon_{o,T}E_{a,\alpha}\right) \tag{12}$$

where $t_{o,T}$ and $\varepsilon_{o,T}$ are fitting coefficients. The effect of temperature on the value of fitting coefficients is shown in **Table 2**.

Since:

$$\left(\frac{d\alpha}{dt}\right)_{v_h} = \left[\left(\frac{\beta_T}{v_h\eta_T}\right)\right]\left[\left(\frac{T-T_o}{v_h\eta_T}\right)\right]^{\beta_T-1} exp\left[-\left(\frac{T-T_o}{v_h\eta_T}\right)\right]^{\beta_T} \tag{13}$$

and

$$\frac{dt}{dE_a} = t_{o,T}\varepsilon_{o,T}\, exp\left(\varepsilon_{o,T}E_{a,\alpha}\right) \tag{14}$$

by applying Eq. (11) $p(E_a)$ was calculated.

Figure 8 shows calculated values of p(E_a).

The $g(E_a)$ under IT has an asymmetric bell shape, invariant with temperature, which clearly shows a maximum of $g(E_a)$ = 0.122 at E_a = 45.8 kJ/mol. The determined temperature dependence of the g(Ea) function indicates that it is right related to the distribution of reactivity of dehydration species in absorbed water at a given type of their activation. The calculated values of $g(E_a)$, according to the shape and the values of the characteristic quantities [40], differ significantly from the values of $g(E_a)$ obtained when calculating the NIT of absorbed water in PAAH. The established difference in the shape and values of the characteristic quantities $g(E_a)$ during non-isothermal and isothermal heating clearly indicates a different mechanism of activation of reaction species during different modes of heating the system.

3.3 Kinetics of nonisothermal dehydration of PAMAH

Kinetics of NIT of PAMAH has been investigated in detail [38]. **Figure 9** shows non-isothermal conversion curves of absorbed water in PAMH at different heating rates.

As in the case of NIT PAAH conversion curves, at all investigated heating rates, they have a complex asymmetric shape. As the heating rate increases, the degree of asymmetry of the conversion curves increases and also the temperature interval within which DH takes place is diminished. Since the conversion curves of DH PAMH cannot be successfully fitted by the most commonly used reaction models characteristic of reactions with the participation of a solid phase [57] and the models of kinetics of DH PAAH shown above. Naya et al. [58] and Cao [59] used the logistic function (Eq. (15)) for the mathematical description of the complex non-isothermal kinetics of polymer degradation:

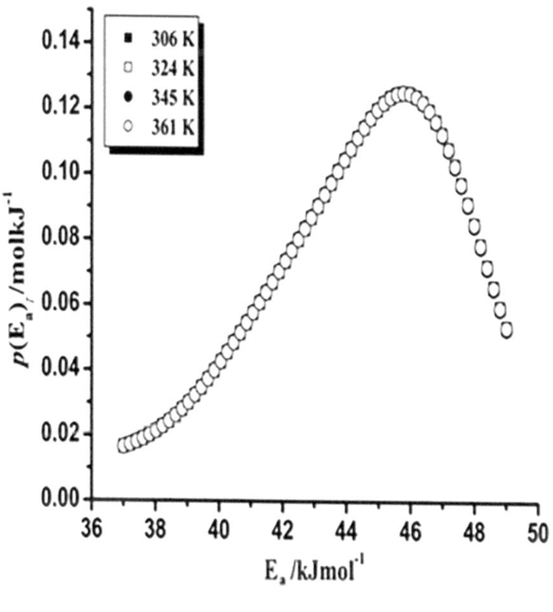

Figure 8.
Effect of temperatures on the shape of the calculated density distribution function of activation energies ($p(E_a)_T$ *[40]*.

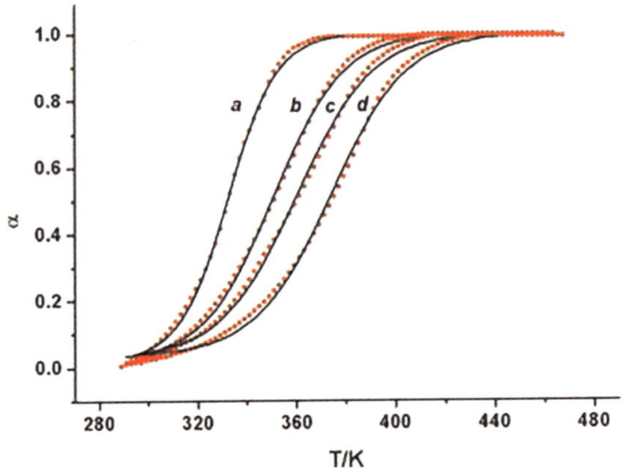

Figure 9.
Conversion curves of NIT of PAMAH at different heating rates (dots) and their mathematical fittings using logistic function (line): (a) v_h = 5 K min^{-1}; (b) v_h = 10 K min^{-1}; (c) v_h = 20 K min^{-1}; (d) v_h = 30 K min^{-1} [38].

Heating rate (K min^{-1})	W	*a*	*b* (K^{-1})	R^2
5	0.98	−33.59	0.10	0.9992
10	0.98	−25.08	0.07	0.9988
20	0.97	−23.70	0.06	0.9984
30	0.96	−24.62	0.06	0.9981

Table 3.
Effect of heating rate on logistic function parameter values and corresponding R^2 values between experimental data and the logistic regression model prediction.

$$\alpha(T)_{v_h} = 1 - \left[\frac{w}{1 + exp\,(a + bT)}\right] \tag{15}$$

where w is experimentally achieved the maximum degree of conversion, and a and b are the parameters of the logistic function.

Bearing that in mind, the experimental DH conversion curves obtained at different heating rates were fitted by Eq. (15). **Figure 9** (solid lines) shows the calculated conversion curves at different heating rates. As can be seen from **Figure 9**, the DH conversion curves of PAMH fitted with the logistic function completely mathematically describe the experimental DH curves. **Table 3** shows the effect of the heating rate on the parameters of the logistic function.

The values of parameters w and b decrease with the increase in heating rate, while the value of parameter a changes complexly with the decrease in heating rate. There is no data in the literature about the physical meaning of applying the logistic function to describe the kinetics of dehydration. Bearing in mind that the logistic function is mathematically very similar to the Prout-Tomkins' eq. [60, 61], which was developed to describe the isothermal degradation of $KMnO_4$. A comparative analysis of the connection between the logistic function and the Prout-Tomkins equation was performed. Based on that analysis, it was concluded that there are functional connections between the parameter of the logistic function (w,b) and the rate constants of nucleus branching (k_b) and nucleus termination (k_t), which are described by expressions (16) and (17).

$$w_{v_h} = \frac{k_{b,v_h}}{k_{T,v_h}} \quad (16)$$

$$b_{v_h} = \frac{k_{b,v_h}}{v_h} \quad (17)$$

Table 4 shows the effect of the heating rate at k_b and k_t.

At all of the investigated heating rates, the value of k_t and k_b increases with the increase in heating rate the values of both k_t and k_b increase. The established functional relationships between the parameters of the logistic function and value of k_t indicate that the logistic function can describe kinetically complex reactions with 3 elementary kinetic stages: nucleation, autocatalysis, and termination.

The established possibility of a mathematical description of the kinetics of NITof PAAGH allows assuming a new model of the mechanism of dehydration of adsorbed water from the hydrogel. According to that model, the DH of absorbed water does not take place by the immediate release of individual water molecules from the absorbed phase but takes place in three well-defined stages: nucleation (formation of clusters of water molecules of critical dimensions), autocatalytic growth of the formed nuclei and termination (decrease in the concentration of nuclei due to the completion process). Bearing in mind the fluctuating structure of the hydrogel, formation of critical water-dehydrating nuclei and their dehydration leads to the collapse of the existing fluctuating structure of the hydrogel. The newly formed fluctuating structure of the hydrogel enables the formation of a large number of dehydration nuclei (autocatalytic growth), which leads to an abrupt acceleration of the dehydration process. At high degrees of dehydration, the rate of dehydration decreases due to the decrease in the number of dehydration nuclei and the transformation of the hydrogel into a xerogel.

3.4 Kinetics of non-isothermal dehydration of PAAGH

Kinetics of non-isothermal dehydration of PAAGH has been investigated by using distributed activation energy model [33]. The TG curves of the NIT of PAAGH are shown in **Figure 10**.

TG curves NIT of PAGH have a complex asymmetric shape (concave down). With an increase in v_h, there is an increase in the degree of symmetry and a widening of the temperature interval in which the DH process takes place. In order to determine the degree of kinetic complexity of dehydration using Vyzovkin's method [62] and Kissinger-Akahira and Sunozze (KAS) method [63, 64], the dependences of E_a and lnA on α were determined. The dependence of E_a on α is shown in **Figure 11** and lnA on α in **Figure 12**.

v_h (K min^{-1})	k_h (min^{-1})	k_T (min^{-1})
5	0.505	0.515
10	0.716	0.731
20	1.316	1.356
30	1.974	2.047

Table 4.
Effect of heating rate at k_b *and* k_t.

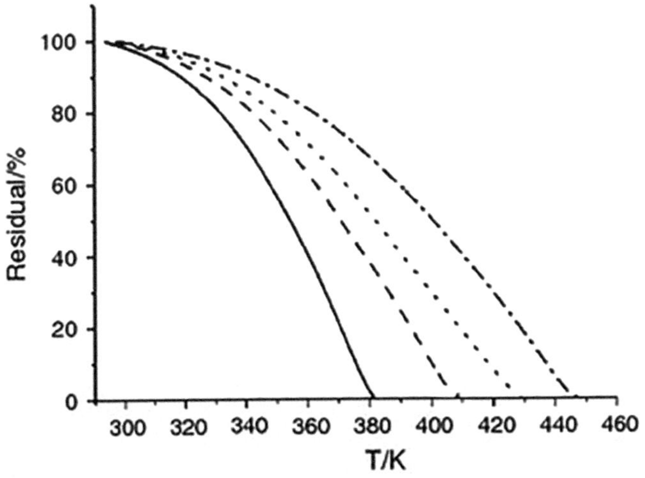

Figure 10.
Thermogravimetric curves of NIT of PAAGH at v_h: 5 K min^{-1} (thick line), 10 K min^{-1} (spaced line), 15 K min^{-1} (dotted line), and 20 K min^{-1} (spaced dotted line) [33].

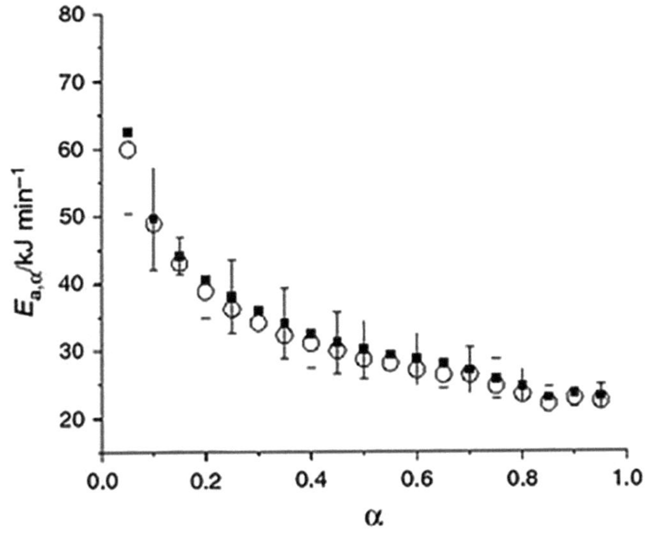

Figure 11.
Dependence of $E_{a,\alpha}$ *on* α *calculated by Vyazovkin's (circle) and KAS (square) methods [33].*

The results shown in **Figures 11** and **12** indicate that both methods of calculating kinetic's parameters lead to identical dependences of E_a and lnA on α (within the limits of experimental error). The complex shape of dependence of E_a and lnA on α confirms the kinetic's complexity of the DH process. On the **Figures 11** and **12**, the 2 characteristic shapes of change of E_a and lnA from α can be noticeably observed within the range $\alpha \leq 0.16$ values of $E_{a,\alpha}$ and lnA_α decrease almost linearly with increasing α. On the contrary, at $\alpha \geq 0.15$, increasing the value of α leads to a slow decrease of E_a from 40.5 to 24 kJ/mol, that is, lnA from 13 to 5 min^{-1}. Between the values $E_{a,\alpha}$ and lnA_α, there is a linear correlation relationship (compensation effect) which is mathematically described by the Eq. (18)

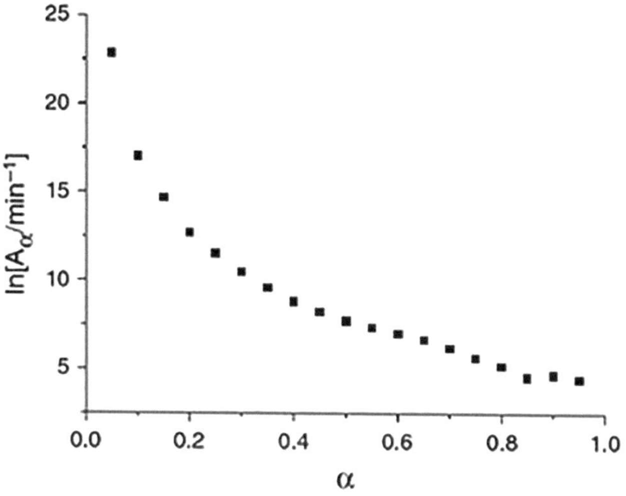

Figure 12.
Dependence of $ln[A_\alpha/min^{-1}]$ *on* α *[33].*

$$ln\,[A_\alpha] = -6.43 + 0.47E_{a,\alpha} \tag{18}$$

In the work of Šimon [65], the so-called single-step approximation for describing the kinetics of chemical reactions and physical-chemical processes that take place in a solid state. In accordance with it, the rate of solid-state reaction can be described by Eq. (19) where $k(T)$ is the reaction rate constant and $f(\alpha)$ is the function that describes the reaction model of the reaction.

$$\frac{d\alpha}{dt} = k(T)f(\alpha) \tag{19}$$

In the case of kinetically complex reactions, Eq. (19) transforms into Eq. (20):

$$\frac{d\alpha}{dt} = A_{\alpha,T}\,exp\left(-\frac{E_{a,\alpha}}{RT_\alpha}\right)f(\alpha)_T \tag{20}$$

If it takes place in NIT conditions, Eq. (20) is transformed into Eq. (21):

$$\left(\frac{d\alpha}{dT}\right)_{v_v} = \frac{A_\alpha\,exp\left(-\frac{E_{a,\alpha}}{RT_\alpha}\right)}{v_v}f(\alpha)_T \tag{21}$$

Eq. (21) enables to calculate the dependence of $k(\alpha)$ and $f(\alpha)$ on temperature based on the dependence of Ea and lnA on α and on the experimentally calculated dependence using the expressions (22) and (23).

$$k(\alpha)_{v_h} = A_{\alpha,v_h}\,exp\left(-\frac{E_{a,\alpha,v_h}}{RT_{\alpha,v_h}}\right) \tag{22}$$

$$f(\alpha)_{v_h} = \frac{v_h\left(\frac{d\alpha}{dT}\right)_{v_h}}{k(\alpha)_{v_h}} \tag{23}$$

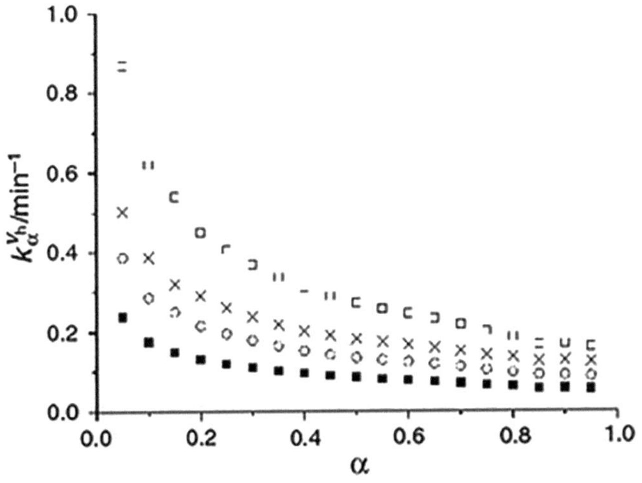

Figure 13.
Dependence of kvhα on α at different v $_h$: 5 K mol^{-1} (filled square), 10 K mol^{-1} (circle), 15 K mol^{-1} (times symbol), 20 K mol^{-1} (empty square) [33].

Figure 13 shows the dependence of k(α) on α.

The values of k(α) nonlinearly decrease with the increase in α, for each of the investigated v_h. At a particular value of α (iso conversion), k(α) increases with an increase in v_h, which designates that k(α) decreases with the increase in temperature. The functional dependence of the reaction model of dehydration on α is shown in **Figure 14**.

On all of the investigated v_h f(α) has a similar shape and gradually increases with the increase of α up to α = 0.85. A further increase in α leads to a sharp decrease in the value of *f*(α). The functional relationship between f(α) and α can be mathematically described by the expression (24).

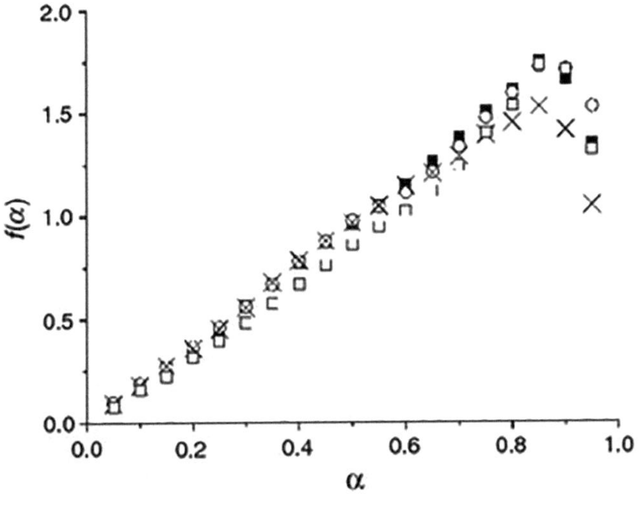

Figure 14.
Function f(α) at different v_h: 5 K mol^{-1} (filled square), 10 K mol^{-1} (circle), 15 K mol^{-1} (times symbol), 20 K mol^{-1} (empty square).

v_h (K min^{-1})	n	m
5	1.97	1.04
10	1.93	1.02
15	1.94	1.02
20	1.84	1.09

Table 5.
Effect of v_h *on values of* m *and* n.

$$f(\alpha)_{v_h} = m_{v_h}\alpha^n \tag{24}$$

Table 5 shows the effects of v_h on m and n, where n and m are the parameters of expression (24).

An increase in v_h leads to a decrease in the value of the parameter n, while the parameter m changes slightly. The independence of the shape of f is $f(\alpha)$ from v_h indicates that the basic cause of the complex nature of dehydration kinetics is the change in k(α) caused by changes in the structure of the absorbed phase caused by DH. The established shape of the dependence of k(α) and f(α) also enables explained the phenomenon of the decrease in the value of v_h DH with T, that is, α. That happens during the investigated non-isothermal dehydration experiments in all cases when the rate of decrease in the value of k(α) is higher than the rate of increase in the value of f(α).

4. Conclusion

The kinetics of DH of all of the investigated hydrogels is of a complex nature. The complex kinetics of dehydration of hydrogels is a consequence of the fluctuating structure of the hydrogel, the phase state of the absorbed water, and the thermal activation of the hydrogel. By applying DAEM, isothermal and non-isothermal kinetics of PAAH dehydration were fully described. Procedures for determining the shape of $g(E_a)$ have been developed. It was shown that the method of activation (isothermal and non-isothermal heating) of the system affects the shape of $g(E_a)$ and the value of the kinetic parameters. The logistic model fully describes the kinetics of NIT dehydration of PAMAH. The functional relationship between the parameters of the logistic function and the constants k_b and k_t was determined. It was shown that the rate of nucleation of dehydration centers is the kinetically limiting stage of PAMAH dehydration rate. By applying the coupled single step-approximation and calculating the iso conversion dependence of Ea and lnA on α, a new kinetic model was developed to describe the complex kinetics of PAAGH dehydration. It was found that the rate constant of dehydration at all investigated v_h decreases with the increase in temperature, while the values of the function parameters of the reaction model increase. The decrease in the value of the dehydration rate constant with temperature is associated with the change in the rate of nucleation.

Acknowledgements

The present investigations were supported by The Ministry of Education, Science and Technological Development of the Republic of Serbia, under Contracts No: 451-03-68/2022-14/200051; 451-03-68/2022-14/200146; 451-03-47/2023-01/200051.

Author details

Jelena D. Jovanović[1*] and Borivoj K. Adnadjević[2]

1 Institute of General and Physical Chemistry, Studentski Trg, Belgrade, Serbia

2 Faculty of Physical Chemistry, University of Belgrade, Studentski Trg, Belgrade, Serbia

*Address all correspondence to: jelenajov2000@yahoo.com

References

[1] Wichterle O, Lim D. Hydrophilic gels for biological use. Nature. 1960; **185**(4706):117-118. DOI: 10.1038/185117a0

[2] Peppas NA, editor. Hydrogels in Medicine and Pharmacy: Fundamentals. Boca Raton: CRC Press; 1986. p. 192. DOI: 10.1201/9780429285097

[3] Omidian H, Park K. Introduction to hydrogels. In: Ottenbrite RM, Park K, Okano T, editors. Biomedical Applications of Hydrogels Handbook. New York, New York, NY: Springer; 2010. pp. 1-16. DOI: 10.1007/978-1-4419-5919-5_1

[4] Peppas NA, Slaughter BV, Kanzelberger MA. 9.20 - Hydrogels. In: Matyjaszewski K, Möller M, editors. Polymer Science: A Comprehensive Reference. Amsterdam: Elsevier; 2012. pp. 385-395. DOI: 10.1016/B978-0-444-53349-4.00226-0

[5] Okay O. General properties of hydrogels. In: Gerlach G, Arndt K-F, editors. Hydrogel Sensors and Actuators: Engineering and Technology. Berlin Heidelberg, Berlin, Heidelberg: Springer; 2010. pp. 1-14. DOI: 10.1007/978-3-540-75645-3_1

[6] Peppas NA, Merrill EW. Poly(vinyl alcohol) hydrogels: Reinforcement of radiation-crosslinked networks by crystallization. Journal of Polymer Science: Polymer Chemistry Edition. 1976;**14**(2):441-457. DOI: 10.1002/pol.1976.170140215

[7] Peppas NA. Hydrogels of Poly(Vinyl Alcohol) and Its Copolymers, Hydrogels in Medicine and Pharmacy: Polymers. Boca Raton: CRC Press Inc; 1986. pp. 1-48

[8] Tanaka T. Collapse of gels and the critical endpoint. Physical Review Letters. 1978;**40**(12):820-823. DOI: 10.1103/PhysRevLett.40.820

[9] Shibayama M, Tanaka T. Volume phase transition and related phenomena of polymer gels. In: Dušek K, editor. Responsive Gels: Volume Transitions I. Berlin Heidelberg, Berlin, Heidelberg: Springer; 1993. pp. 1-62. DOI: 10.1007/3-540-56791-7_1

[10] Peppas NA, Bures P, Leobandung W, Ichikawa H. Hydrogels in pharmaceutical formulations. European Journal of Pharmaceutics and Biopharmaceutics. 2000;**50**(1):27-46. DOI: 10.1016/S0939-6411(00)00090-4

[11] Lin C-C, Metters AT. Hydrogels in controlled release formulations: Network design and mathematical modeling. Advanced Drug Delivery Reviews. 2006; **58**(12):1379-1408. DOI: 10.1016/j.addr.2006.09.004

[12] Blanchard J. Controlled Drug Delivery: Challenges and Strategies Edited by Kinam Park (Purdue University). American Chemical Society: Washington, DC. 1997. xvii + 629 pp. $145.95. ISBN 0-8412-3418-3. Journal of the American Chemical Society. 1998;**120**(18):4554-4555. DOI: 10.1021/ja9756228

[13] Peppas NA. Hydrogels and drug delivery. Current Opinion in Colloid & Interface Science. 1997;**2**(5):531-537. DOI: 10.1016/S1359-0294(97)80103-3

[14] Alma TB, Chigozie AN, Emmanuel MB. Recent advances in biosensing in tissue engineering and regenerative medicine. In: Vahid A, Selcan K, editors. Biosignal Processing. Rijeka: IntechOpen; 2022. DOI: 10.5772/intechopen.104922

[15] Caló E, Khutoryanskiy VV. Biomedical applications of hydrogels: A review of patents and commercial products. European Polymer Journal. 2015;**65**:252-267. DOI: 10.1016/j.eurpolymj.2014.11.024

[16] Hunt JA, Chen R, van Veen T, Bryan N. Hydrogels for tissue engineering and regenerative medicine. Journal of Materials Chemistry B. 2014;**2**(33): 5319-5338. DOI: 10.1039/c4tb00775a

[17] Guo Y, Bae J, Fang Z, Li P, Zhao F, Yu G. Hydrogels and hydrogel-derived materials for energy and water sustainability. Chemical Reviews. 2020; **120**(15):7642-7707. DOI: 10.1021/acs.chemrev.0c00345

[18] Richbourg NR, Peppas NA. The swollen polymer network hypothesis: Quantitative models of hydrogel swelling, stiffness, and solute transport. Progress in Polymer Science. 2020;**105**: 101243. DOI: 10.1016/j.progpolymsci.2020.101243

[19] Jovanovic J, Adnadjevic B. The effect of primary structural parameters of poly (methacrylic acid) xerogels on the kinetics of swelling. Journal of Applied Polymer Science. 2013;**127**(5):3550-3559. DOI: 10.1002/app.37706

[20] Chamerski K, Korzekwa W, Filipecki J, Shpotyuk O, Stopa M, Jeleń P, et al. Nanoscale observation of dehydration process in PHEMA hydrogel structure. Nanoscale Research Letters. 2017;**12**(1):303. DOI: 10.1186/s11671-017-2055-3

[21] Krysztofiak K, Szyczewski A. Study of dehydration and water states in new and worn soft contact lens materials. Optica Applicata. 2014;**44**:237-250

[22] Fonn D, Situ P, Simpson T. Hydrogel lens dehydration and subjective comfort and dryness ratings in symptomatic and asymptomatic contact lens wearers. Optometry and Vision Science. 1999; **76**(10):700-704

[23] Ramamoorthy P, Sinnott LT, Nichols JJ. Treatment, material, care, and patient-related factors in contact lens-related dry eye. Optometry and Vision Science. 2008;**85**(8)

[24] Krysztofiak K, Płucisz M, Szyczewski A. The influence of wearing on water states and dehydration of silicone-hydrogel contact lenses. Engineering of Biomaterials/Inżynieria Biomateriałów. 2012;**115**:18-25

[25] Sirousazar M, Kokabi M, Yari M. Mass transfer during the pre-usage dehydration of polyvinyl alcohol hydrogel wound dressings. Iranian Journal of Pharmaceutical Sciences. 2008;**4**(1):51-56

[26] Hawlader MNA, Ho JC, Qing Z. A mathematical model for drying of shrinking materials. Drying Technology. 1999;**17**(1–2):27-47. DOI: 10.1080/07373939908917517

[27] Roques MA, Zagrouba F, Sobral PDOA. Modelisation principles for drying of gels. Drying Technology. 1994;**12**(6):1245-1262. DOI: 10.1080/07373939408961004

[28] Kemp IC, Fyhr BC, Laurent S, Roques MA, Groenewold CE, Tsotsas E, et al. Methods for processing experimental drying kinetics data. Drying Technology. 2001;**19**(1):15-34. DOI: 10.1081/drt-100001350

[29] Pekcan Ö, Yilmaz Y. Flourescence method to study gelation swelling and drying processes in gels formed by solution free radical compolymerization. In: Zrínyi M, editor. Gels. Darmstadt: Steinkopff; 1996. pp. 89-97

[30] Tarı Ö, Pekcan Ö. Study of drying of κ-carrageenan gel at various temperatures using a fluorescence technique. Drying Technology. 2007;**26**(1):101-107. DOI: 10.1080/07373930701781728

[31] Evingür GA, AktaŞ DK, Pekcan Ö. In situ steady state fluorescence (SSF) technique to study drying of PAAm hydrogels made of various cross-linker contents. Chemical Engineering and Processing: Process Intensification. 2009;**48**(2):600-605. DOI: 10.1016/j.cep.2008.07.003

[32] Janković B, Adnađević B, Jovanović J. Non-isothermal kinetics of dehydration of equilibrium swollen poly(acrylic acid) hydrogel. Journal of Thermal Analysis and Calorimetry. 2005;**82**(1):7-13. DOI: 10.1007/s10973-005-0885-1

[33] Stankovic B, Jovanovic J, Ostojic S, Adnadjevic B. Kinetic analysis of non-isothermal dehydration of poly(acrylic acid)-g-gelatin hydrogel using distributed activation energy model. Journal of Thermal Analysis and Calorimetry. 2017;**129**(1):541-551. DOI: 10.1007/s10973-017-6180-0

[34] Sovizi MR, Fakhrpour G, Bagheri S, Rezanejade Bardajee G. Non-isothermal dehydration kinetic study of a new swollen biopolymer silver nanocomposite hydrogel. Journal of Thermal Analysis and Calorimetry. 2015;**121**(3):1383-1391. DOI: 10.1007/s10973-015-4639-4

[35] Liu L, Wang B, Gao Y, Bai TC. Chitosan fibers enhanced gellan gum hydrogels with superior mechanical properties and water-holding capacity. Carbohydrate Polymers. 2013;**97**(1): 152-158. DOI: 10.1016/j.carbpol.2013.04.043

[36] Liu L, Wang B, Bai TC, Dong B. Thermal behavior and properties of chitosan fibers enhanced polysaccharide hydrogels. Thermochimica Acta. 2014; **583**:8-14. DOI: 10.1016/j.tca.2014.03.008

[37] Ma Y, Bai T, Wang F. The physical and chemical properties of the polyvinylalcohol/polyvinylpyrrolidone/hydroxyapatite composite hydrogel. Materials Science and Engineering: C. 2016;**59**:948-957. DOI: 10.1016/j.msec.2015.10.081

[38] Adnadjevic B, Tasic G, Jovanovic J. Kinetic of non-isothermal dehydration of equilibrium swollen poly(acrylic acid-co-methacrylic acid) hydrogel. Thermochimica Acta. 2011;**512**(1): 157-162. DOI: 10.1016/j.tca.2010.09.019

[39] Adnađević B, Janković B, Kolar-Anić L, Minić D. Normalized Weibull distribution function for modelling the kinetics of non-isothermal dehydration of equilibrium swollen poly(acrylic acid) hydrogel. Chemical Engineering Journal. 2007;**130**(1):11-17. DOI: 10.1016/j.cej.2006.11.007

[40] Adnađević B, Janković B, Kolar-Anić L, Jovanović J. Application of the Weibull distribution function for modeling the kinetics of isothermal dehydration of equilibrium swollen poly (acrylic acid) hydrogel. Reactive and Functional Polymers. 2009;**69**(3): 151-158. DOI: 10.1016/j.reactfunctpolym.2008.12.011

[41] Adnadjevic B, Jovanovic J, Micic U. The kinetics of isothermal dehydration of equilibrium swollen hydrogel of poly (acrylic-co-methacrylic acid). Hemijska Industrija. 2009;**63**(5):585-591. DOI: 10.2298/hemind0905585a

[42] Janković B, Adnađević B, Jovanović J. The comparative kinetic study of non-isothermal and isothermal dehydration of swollen poly(acrylic acid) hydrogel using the Weibull probability function. Chemical

Engineering Research and Design. 2011; **89**(4):373-383. DOI: 10.1016/j.cherd.2010.09.001

[43] Gm B, Yv Z. Kurs fiziki polimerov (Course of Polymer Physics). 1st ed. Leningrad: Khimiya; 1976

[44] Potkonjak B, Jovanović J, Stanković B, Ostojić S, Adnadjević B. Comparative analyses on isothermal kinetics of water evaporation and hydrogel dehydration by a novel nucleation kinetics model. Chemical Engineering Research and Design. 2015; **100**:323-330. DOI: 10.1016/j.cherd.2015.05.032

[45] Bellich B, Borgogna M, Cok M, Cesàro A. Release properties of hydrogels: Water evaporation from alginate gel beads. Food Biophysics. 2011;**6**(2):259-266. DOI: 10.1007/s11483-011-9206-3

[46] Bellich B, Borgogna M, Cok M, Cesàro A. Water evaporation from gel beads: A calorimetric approach to hydrogel matrix release properties. Journal of Thermal Analysis and Calorimetry. 2011;**103**(1):81-88. DOI: 10.1007/s10973-010-1170-5

[47] Heyd R, Rampino A, Bellich B, Elisei E, Cesàro A, Saboungi M-L. Isothermal dehydration of thin films of water and sugar solutions. The Journal of Chemical Physics. 2014;**140**(12):124701. DOI: 10.1063/1.4868558

[48] Stanković B, Jovanović J, Adnađević B. The kinetics of non-isothermal dehydration of equilibrium swollen Ca-alginate hydrogel. Journal of Thermal Analysis and Calorimetry. 2020;**142**(5):2123-2129. DOI: 10.1007/s10973-020-10020-6

[49] Adnadjević B, Gigov M, Jovanović J. Comparative analyses on isothermal kinetics of water evaporation and PAAG hydrogel dehydration under the microwave heating conditions. Chemical Engineering Research and Design. 2017; **122**:113-120. DOI: 10.1016/j.cherd.2017.04.009

[50] Jovanović J, Adnadjević B, Ostojić S, Kićanović M. An investigation of the dehydration of superabsorbing polyacrylic hydrogels. Materials Science Forum. 2004; **453–454**:543-548. DOI: 10.4028/www.scientific.net/MSF.453-454.543

[51] Adnadjevic B, Jovanovic J. A comparative kinetics study of isothermal drug release from poly(acrylic acid) and poly(acrylic-co-methacrylic acid) hydrogels. Colloids and Surfaces B: Biointerfaces. 2009;**69**(1):31-42. DOI: 10.1016/j.colsurfb.2008.10.018

[52] Friedman HL. Kinetics of thermal degradation of char-forming plastics from thermogravimetry. Application to a phenolic plastic. Journal of Polymer Science. 1964;**6**(1):183-195

[53] Lakshmanan CC, White N. A new distributed activation energy model using Weibull distribution for the representation of complex kinetics. Energy & Fuels. 1994;**8**(6):1158-1167. DOI: 10.1021/ef00048a001

[54] Burnham AK, Weese RK, Weeks BL. A distributed activation energy model of thermodynamically inhibited nucleation and growth reactions and its application to the $\beta-\delta$ phase transition of HMX. The Journal of Physical Chemistry B. 2004; **108**(50):19432-19441. DOI: 10.1021/jp0483167

[55] Vyazovkin S. Computational aspects of kinetic analysis.: Part C. The ICTAC kinetics project—The light at the end of the tunnel? Thermochimica Acta. 2000; **355**(1):155-163. DOI: 10.1016/S0040-6031(00)00445-7

[56] Weibull W. A statistical distribution function of wide applicability. Journal of Applied Mechanics. 1951;**18**:293-297

[57] Khawam A, Flanagan DR. Solid-state kinetic models: Basics and mathematical fundamentals. The Journal of Physical Chemistry B. 2006;**110**(35):17315-17328. DOI: 10.1021/jp062746a

[58] Naya S, Cao R, Artiaga R. Local polynomial estimation of TGA derivatives using logistic regression for pilot bandwidth selection. Thermochimica Acta. 2003;**406**(1–2): 177-183

[59] Cao R, Naya S, Artiaga R, Garcia A, Varela A. Logistic approach to polymer degradation in dynamic TGA. Polymer Degradation and Stability. 2004;**85**(12): 667-674. DOI: 10.1016/j.polymde gradstab

[60] Prout E, Tompkins FC. The thermal decomposition of silver permanganate. Transactions of the Faraday Society. 1946;**42**:468-472

[61] Omidian H, Hashemi S, Sammes P, Meldrum I. A model for the swelling of superabsorbent polymers. Polymer. 1998;**39**(26):6697-6704

[62] Vyazovkin S, Dollimore D. Linear and nonlinear procedures in isoconversional computations of the activation energy of nonisothermal reactions in solids. Journal of Chemical Information and Computer Sciences. 1996;**36**:42-45

[63] Kissinger HE. Reaction kinetics in differential thermal analysis. Analytical Chemistry. 1957; **29**(11):1702-1706. DOI: 10.1021/ ac60131a045

[64] Akahira T, Sunose T. Method of determining activation deterioration constant of electrical insulating materials. Research Report Chiba Institute of Technology (Sci Technol). 1971;**16**(1971):22-31

[65] Šimon P. Considerations on the single-step kinetics approximation. Journal of Thermal Analysis and Calorimetry. 2005;**82**(3):651-657. DOI: 10.1007/s10973-005-0945-6